# GLOBAL INSANITY

## How *Homo sapiens* Lost Touch with Reality while Transforming the World

# GLOBAL INSANITY

## How *Homo sapiens* Lost Touch with Reality while Transforming the World

James A. Coffman & Donald C. Mikulecky

**EMERGENT**™
PUBLICATIONS

3810 N 188th Ave
Litchfield Park, AZ 85340

*Global Insanity:*
*How Homo sapiens Lost Touch with Reality while Transforming the World*

Written by: James A. Coffman & Donald C. Mikulecky

Library of Congress Control Number: 2012954070

ISBN: 978-1-938158-04-9

Copyright © 2012 3810 N 188th Ave, Litchfield Park, AZ 85340, USA

All rights reserved. No part of this publication may be reproduced, stored on a retrieval system, or transmitted, in any form or by any means, electronic, mechanical, photocopying, microfilming, recording or otherwise, without written permission from the publisher.

Printed in the United States of America

# ABOUT THE AUTHORS

**James A. Coffman** is an Associate Professor at the Mount Desert Island Biological Laboratory in Salisbury Cove, Maine. He earned a Ph.D. in Zoology in 1990 at Duke University and did postdoctoral training with Eric Davidson at the California Institute of Technology. As a developmental biologist he has used sea urchins as an experimental model to investigate how ontogeny is controlled by genetic information, cell physiology and environmental conditions. He is an author of numerous refereed journal articles. Over the past decade he has been exploring the idea that evolutionary change in complex systems is progressively constrained by their development.

**Donald C. Mikulecky** is a Senior Fellow in the Virginia Commonwealth University Center for the Study of Biological Complexity. He earned a Ph.D. in Physiology in 1963 at the University of Chicago and did postdoctoral training with the late Aaron (Katzir) Katchalsky in non-equilibrium thermodynamics and membrane biophysics at the Weizmann Institute in Israel. As a theoretical biologist he has specialized in creating and testing mathematical models of biological systems and processes. He is the author of the book *The Application of Network Thermodynamics to Problems in Biomedical Engineering* and scores of refereed journal articles. In the last twenty or thirty years he has been exploring the complexity theory started by the late Robert Rosen.

# ACKNOWLEDGEMENTS

This book grew out of our convergent interests in complexity science and the work of Robert Rosen, and an e-mail dialog that began when one of us (JAC) complimented the other (DCM) on a paper (Mikulecky, 2011) that he had recently posted on his webpage. We have never met in person, but it became clear early on in our collaboration that we are kindred spirits who think about and relate to the world in much the same way. We can thank digital computer technology—perhaps the crowning achievement of the Industrial Revolution—for catalyzing (significantly lowering the 'potential barrier' of geography that would otherwise impede) discourse in general, and this work in particular.

We thank Bob Ulanowicz, Stan Salthe, A. Louie, Alicia Juarrero, and Jaime Gomez for providing helpful comments on an early draft of this essay, and Geoffrey Klempner for publishing an early version of Chapter 2 in his e-journal *Philosophy Pathways*.

While we have tried to cite those whose work directly informed the ideas that we articulate, we have kept the number of citations to a minimum in order to maximize readability. In many cases this was done by citing secondary sources containing more extensive bibliographies, to which the reader is referred. We apologize to all those who have indirectly contributed to our thinking or who have expressed similar ideas, who we did not explicitly acknowledge as sources.

Two important sources who we did not cite (owing to our not having actually read their original works), yet whose ideas permeate much of this essay, are the American logician/philosopher/mathematician/scientist Charles Sanders Peirce—a founder of semiotics (the study of how communication occurs by way of signs and symbols)—and the Swiss psychologist/psychiatrist Carl Gustav Jung. The erudite reader will recognize their influence in several of the themes that we develop.

Special thanks go to Sandra Coffman, Ph.D., whose keen insight into human psychology, and how physiology and social dynamics constrain and entrain personality, informed much of what is written or implied here about those subjects.

James A. Coffman and Donald C. Mikulecky
Mount Desert Island, Maine and Mathews, Virginia
September, 2012.

# CONTENTS

**Introduction**
**OUR THESIS, AND WHAT WE HOPE TO ACHIEVE** ..................... 1

**Chapter 1**
**SYSTEM EARTH IN THE ANTHROPOCENE** ............................... 9

**Chapter 2**
**OF METAPHORS, METAPHYSICS, AND MATH:**
**A MYTHOLOGY OF MECHANISMS** ......................................... 17
    Plato and Aristotle ............................................................... 28
    Bacon, Descartes and Newton .............................................. 30

**CHAPTER 3**
**ROBERT ROSEN'S INSIGHT INTO LIFE ITSELF** ....................... 37
    Teleonomic Constraint: Evolved Functionality ....................... 39
    Teleomatic Change: Going with the Flow .............................. 43
    The Modeling Relation ......................................................... 46
    Simple versus Complex Systems ........................................... 49
    On the Generality of Life ...................................................... 55

**CHAPTER 4**
**THE LOGIC OF DEVELOPMENT** ............................................. 59
    Growing (Inter)Dependency .................................................. 65
    Dialogic ................................................................................ 70
    Subjective Individuation ....................................................... 75
    Entrenchment ...................................................................... 80

**CHAPTER 5**
**METABOLISM AND REPAIR IN THE GLOBAL ECONOMY** ....... 87

**Chapter 6**
**RUNNING ON EMPTY** ......................................................... 101

**Chapter 7**
**WHAT CAN BE DONE?** ........................................................ 115

**REFERENCES** ..................................................................... 139

# PREFACE

Sometimes things won't change for the better unless you bite the hand that feeds you.

In many ways that is what this book is about.

We expect that many readers will agree that the present outlook is not good; and hence, that "change for the better" is needed. The problem however is that such change will not happen unless a significant number of us begin doing things that work against our own short-term interests.

Our writing and publishing this book may well be one such thing.

---

To work against your own short-term interests requires that you acknowledge that those interests are harmful or *unhealthy*. So it is reasonable to ask why more people do not acknowledge and act on the increasingly serious health problems that confront us all as a result of anthropogenic environmental degradation. The answer to this is, in a word, complex.

The efforts of science to speak to the condition of the planet are often met with hostility, to say the least. One working scientist who has been subjected to a living hell merely for doing his job is climatologist Michael Mann. A quote from his recent book gives a flavor of what is at stake:

> *We look back now with revulsion at the corporate CEOs, representatives, lobbyists, and scientists-for-hire who knowingly ensured the suffering and mortality of millions by hiding their knowledge of tobacco smoking's ill effects for the sake of short term corporate profits. Will we hold those who have funded or otherwise participated in the fraudulent denial of climate change similarly accountable—those individuals and groups who both made and took corporate payoffs for knowingly lying about the threat climate change posed to humanity, those who willfully have led the public and policy makers astray, and those politicians and media figures who have sought to intimidate climate scientists using McCarthyite tactics?* (Mann, 2012)

Two things need to be pointed out in preparation for what we are offering here. First, the issues described by Mann are not isolated. The entire spectrum and character of political and civic life is subject to powerful economic forces, and while those forces may on the surface reduce to simple greed, below the surface are undercurrents that are not at all that simple, emanating from among other things our animal nature, human psychology and the historical origins of our culture. The second point concerns why political arguments cannot be won with empirically demonstrable facts and logic. We will attempt to address these issues in what follows.

---

The "hand that feeds" the two of us is academic science. We expect that our message will annoy many of our scientific peers.

Why would that work against our own short-term interests?

Because in the game of science reputation is everything.

If a scientist gets a reputation for espousing ideas that are widely viewed as 'unscientific', then he or she may find it harder to get funded. A scientist's funding—and indeed career—depends entirely on the esteem of his or her peers. And to an extent this is well and good, because there are plenty of crackpot ideas that have no basis in reality.

But it is also the case that revolutionary ideas often appear as 'crackpot' to established ways of thinking. As much as we may try to pretend otherwise, scientists are people, subject to the full spectrum of human foibles. And no one likes to consider the possibility (much less to be told) that the way they think about the world is wrong.

And that is what we contend: that despite its awesome transformative power, reductionist science paints—and draws humanity inexorably into—an unrealistic picture of the world.

---

One of the least acknowledged aspects of 'real world' science is the fact that it has always needed a patron within established society. In the early 1960s this took on a new character after the USSR launched Sputnik, which brought science in the United States more deeply into the Cold War. Money poured into education, science and technology, enabling people who would have otherwise been excluded from our ranks—one of us (DCM) for example—to gain entry. The clear surge in science directed by government policy at that moment in history is revealing of the more subtle background of social control that has always been there. What is most difficult to tease out of all this is the way the culture itself has been molded by that atmosphere, and what role that cultural undercurrent plays in what we are going to talk about.

This brings us to an aspect of the reductionist paradigm that is less obvious but perhaps more insidious. As noted above scientific knowledge can pose a threat to powerful economic interests. At the same time the scientifically established reality of evolution is still widely denied by those who prefer to attribute life to the *super*natural. Such folks have no restrictions on rationalizations that keep their stories in line with scientific observations, and they are troubled by the fact that society abounds with contradictory information. This may seem innocent enough, but it represents an unacknowledged political influence buoying many of the sophistic attempts to discredit the results of science.

We will therefore be talking about the nature of knowledge itself, and about the unwitting cognitive-cultural collusion between reductionist science and religion, with the creation myth of fundamentalist Christians being paradoxically part and parcel with the scientifically-informed worldview of Western civilization. In this myth original sin resulted in the fall of all of creation. What better rationalization can be found for human attempts to manipulate and control nature? As arrogant as that rationalization sounds it clearly influences our attitude toward the world and how we as a civilization treat it.

So in a nutshell, we will be taking the radical position that reductionism and religion, to the extent that they represent different forms of *literalism*, work together to perpetuate the Global Economic system by creating a cultural milieu in which people are easily manipulated by (among other things) the economic powers-that-be. The stage is thus set for attacks on science, such as those of coal and oil companies who set up pseudo-scientific organizations and hire scientists to deny the results of climate science. Similarly, when drug companies, which are financially motivated to prevent preventative medicine, manipulate science to their advantage we are again looking at a well-financed undermining of science itself.

---

Many will view this essay itself as being an attack on science. But it is not; if anything it is our attempt to *revive* science. We are scientists and we love science. Our critique is not of science itself, but of a specific overdeveloped strain of science known as reductionism. We contend that the unfettered ascendancy of reductionism by way of the Industrial Revolution explains the existential crisis that confronts the contemporary world. The reason for this is that reductionist science has completely misconceived life, and hence the meaning of *health*.

The Latin word for health is *sanitas*, from whence comes the word 'sanity'. The ill health of the Westernized world—the *Global Insanity* in the title of this book—manifests in myriad ways. Examples include the unmitigated destruction of the environment for the sake of the Economy, despite the widely acknowledged fact that this will only lead to untold suffering and quite possibly the demise of civilization; endless wars that kill millions of innocents and cause widespread suffering and health problems, rationalized by transparent lies; hypocrisy as the social norm, manifesting as total contradiction between what most people (across the political spectrum) say and what they do; widespread denial of empirically verified facts such as biological evolution and climate change; near universal misattribution of socioeconomic problems to a scapegoat, manifesting a fear of the 'bogeyman' *other*, be it real (e.g., worldviews, belief systems, or political parties other than one's own) or superstitious (e.g., Satan); and here in the United States, a culture of paranoia, epitomized by a hugely over-bloated defense budget and the National Rifle Association, which works to ensure that everyone not only has the right to arm themselves to the gills with high-power assault weapons, but also easy access to them. All of these examples (and the list goes on and on) present a disconnection from reality that is injurious to the health of humanity, and are thus symptomatic of Global Insanity. We contend that the disconnection is engendered by the mistaken yet widely-held 'scientific' belief that the living world, including our own species, is a machine (be it evolved or created by 'God'), unworthy of our empathy, which can be adequately if not fully understood through objectification.

The implications of our thesis are far-reaching, and if taken seriously, quite unsettling—particularly for anyone accustomed to thinking that mechanisms are the be-all and end-all of causality in nature.

Such folks do not take kindly to the idea that mechanisms are naught but a means to an end.

They also react strongly to the idea that objectivity is an ideological *myth*.

So be it. What we have to say needs to be said, because if people don't start seeing the world differently, things will not change for the better.

When you come to see the world as we have, there is no going back.

# INTRODUCTION

## OUR THESIS, AND WHAT WE HOPE TO ACHIEVE

The thesis of this essay is that Western science has misconceived life. As a consequence, civilized humanity, by way of its scientifically informed industrial economy cum existential nihilism cum retreat into fantasy, is destroying the biosphere—and hence itself.

The misconception is that life is engendered and fully explained by *mechanisms*.

This is absurd. In biology anything that can be construed as a mechanism can also be logically construed as having a purpose. Means imply ends, and are thus meaningful. Life is neither created by mechanisms, nor an emergent property thereof: to the contrary, mechanisms, to the extent that they perform useful work, are created by living systems in the service of life. Wherever they exist, they do so *in order to* realize some subjective goal.

And at some level, all biologists know this to be true.

Hence, when you attend a lecture by a cell biologist you will often hear a story about what a particular cell is 'trying' to do in a given situation, or how it 'knows' what it must do in order to achieve its 'goal'. The lecturer will typically apologize for 'anthropomorphizing', and the audience will understand this to mean that the speaker does not really mean that the cell in question actually *tries* to do anything—rather, it simply behaves deterministically, and hence predictably, by virtue of molecular mechanisms, which creates the illusion of goal-directed behavior. The anthropomorphic description is thus rationalized as nothing more than metaphorical 'shorthand' necessitated by the limitations of human language.

We are not the first to point out that such rationalizations are problematic (see Juarrero, 1999; Mathews, 2003; Deacon, 2011; and Powell, 2012 for recent critiques). But it still bears asking: what motivates the mental gymnastics that go into making them? We will attempt to answer that question in this essay. For the moment suffice it to say that historically science became committed to the simplifying assumption that, within the material universe, subjectivity and intentional agency are uniquely human attributes. This served to placate clerical authority and allowed science to proceed unfettered by treating everything in the living world as an object—including, ultimately, humanity itself. The ensuing successes of science made it easy to forget that the initial simplifying assumption was merely that, and not something that could be logically or empirically substantiated.

The assumption that subjective agency is uniquely human, and the scientific misconception of life that it engendered, is cognitively dissonant, implying as it does that life itself is without purpose, meaning or intrinsic value. This nihilistic point-of-view is nonetheless passionately promulgated by professional skeptics, who profess that it must be true because science trumps 'mere' belief.

As if it was possible for any point-of-view to exist independently of belief.

It is not. It is nevertheless quite difficult to point out that science itself is a belief structure, for to do so is to invite ridicule and outright dismissal, without being heard.

Is it any wonder that so many people view science with distrust and suspicion?

But *science* is one thing, and *scientism* is another. And it is important to distinguish between the two.

Science is nothing more than a human expression of life itself. Like (and with) life, science evolves. Ideas, like species (the words were originally synonymous) come and go, subject to the natural selection of scientific discourse. This discourse works best with a minimal number of assumptions, and it discards those that, like the assumptions of geocentricity and special creation, become untenable in light of new ideas that better fit reality.

Scientism on the other hand is a system of belief that, like any other system of belief, is motivated by the psychological need for certitude—what Julian Jaynes described as "the search for an innocence of certainty among the mythologies of facts". Scientism wields skepticism not in the service of seeking truth, but merely to cast doubt on any belief that is not open to scientific inquiry (that is, any belief that is not empirically or mathematically testable), or, more commonly, any belief that runs counter to the current scientific view of reality. But it stops short of casting its skeptical eye on the untested assumptions of science itself, because that would undermine the smug certitude that it serves to reinforce.

Scientific knowledge begets scientism because science deals with two distinct aspects of reality: *what it is* (i.e., the (f)actual world), and *how*, both through history and in terms of perceived meaning, *we have come to think about it*. These two aspects are forever entangled in a Gordian knot of what we (think we) know to be true. While science works quite well to increasingly bring the world to light, this necessarily occurs via knowledge, which entails, and cannot possibly be divorced from, language-based thought and interpretation. All knowledge therefore has, and always will have, a subjective aspect, despite the idealistic aspirations of science to eliminate it.

Consider two examples: gravity and evolution. Both are empirical facts descriptive of the actual world. But science seeks not just to describe, but also to explain, which requires creative thinking, the formation of *ideas* ('idealization'). Moreover, factual description itself becomes enabled by explanation. So thanks to the creative genius of Einstein, *gravity* refers not only to the empirical fact that things fall to earth, but also to its idealized explanation involving the curvature of spacetime by mass. Thanks to the creative genius of Darwin *evolution* refers not only to the empirical fact that the living world has arrived at its present state through an eons-long historical process of change involving transformation of distinct life forms from one to another, but also to its idealized explanation involving descent with modification via natural selection. In both cases recognition of the fact preceded the scientific explanation, and in both cases the explanations brought the facts themselves into sharper relief, while also leading to the discoveries of new, previously unknown facts.

But scientism, much like religious literalism, fails to distinguish between the facts of the matter and their idealized explanations. It does so because it is intimately tied to reductionism, a way of thinking that conflates fact with explanation by assuming that both, to the extent that they refer to reality, refer to nothing more than mindless *mechanisms* constructed from the 'bottom up' via the interactions of *objects*. The assumption that life is entirely attributable to mechanistically interacting objects is untenable however, because it occludes context, which is what affords purpose and meaning, as well as value. Owing to this occlusion, the mechanistic *ideal*, rather than the real world, became the center of scientific attention—a sublimely ironic development, given that the ostensible purpose of science is to discern reality.

Scientism encourages us to believe that technology—the fruit of the mechanistic worldview—is the savior of humanity: keep the faith, and all will be well.

But all is not well, and technological solutions always seem to create at least as many problems as they solve. Reality increasingly belies the belief: amidst all our technological wonders, the health and wellbeing of the living planet are not improving; rather, they are in precipitous decline.

There is no denying that technology saved our ancestors from many existential threats, and that it continues to benefit humanity in many ways. The problem with ignoring context however is that doing so hides many of the costs of technological progress, resulting in a progressive loss of regenerative capacity that goes unnoticed until it is too late. At that point collapse becomes unavoidable. This too is a lesson of the living world, for anyone who is paying attention. Much as we try to escape it, death is a reality of life.

What we articulate in the pages that follow is a decidedly non-mechanistic, phenomenological view of life. It is nevertheless a fully naturalistic perspective informed by science, with no recourse to anything mystical or supernatural. What differentiates this view from the standard reductionist view is the insight it grants into the nature of complexity, which reveals fundamental limits on knowledge. Knowledge is seen to be acquired through models, which *must* oversimplify the actual world because they are *embodied* as structurally and functionally constrained thermodynamic systems that develop into existence—meaning that they come to be via arbitrary choices that realize some potentialities at the expense of others. The result is a reinforcement of specific dependencies in the service of some specific, flow-directed end, and a concomitant loss of potential or capacity (i.e., degrees of freedom) for realizing alternative ends. Development is thus seen to be a causal process that progressively channels energy flow by decreasing resistance along some paths of potential while increasing resistance along others.

But resistance causes stress, and life, as a mediator of flow, seeks to eliminate stress however it can. Life strives to be free. And yet, to do what it needs to do (give way to flow) it must self-organize, which erects developmental constraints. The result is a yin and yang dialectic that can never be resolved. The only way life can be free is through death.

And there's the rub. For one reason or another, the human world has built a psychological wall around our fear of death. It is the same wall that separates each of us as individuals from the rest of the world, a wall that reinforces the illusion that we are autonomous agents, somehow separate from an environment that exists 'out there'. But to see that the wall itself is but an illusion all you have to do is ask yourself: where do you draw the line between yourself and the rest of the world, between what is 'internal' and 'external' to you?

On close examination you find that that is not an easy question to answer. For, where do you draw the line when 'you' are but an embryo? Would you draw it at the boundary of the inner cell mass that will eventually become your body (or that will become both you and your identical twin if that mass of cells happens to split in two), or at the expanding edge of the trophectoderm that interpenetrates with the lining of your mother's womb to form your/her placenta? And even after you are born and the cord is cut, where do you draw the line as an infant? Clearly the ability, motivation, and choice of where to draw the line are developed, and this is dependent upon development of individual ego. But the latter is a psychological phenomenon, not a strictly material one: it is possible to have a fully developed body and an undeveloped ego (or even multiple egos at different stages of development, as in 'dissociative identity disorder'!). The perception of a boundary between self and non-self is thus subjective and entirely dependent on psychological development.

Some will argue that objective individuality originates at conception with the unique combination of genes created by union of sperm and egg. But the genetic 'individual' does not arise de novo, but rather is assembled through the recombination of assorted bits of information, DNA sequences that have been passed to us, in many ways unchanged, from remote ancestors that did not look anything like us. Moreover, the individuality of an organism doesn't strictly correlate with genetic identity, as shown by the empirical facts of genetically identical twins on the one hand and genetic mosaics (single individuals that develop from the fusion of two or more genetically distinct embryos) on the other. So why draw the line at conception? We are all related and part of a continuous living lineage; life on earth is in fact one *being*, each creature a branch of a four dimensional tree (or more accurately, intertwining bush) whose roots extend into the primordial soup. Genomes, like fossils, are nothing more than records of biological information, and since information is meaningful only via interpretation, drawing the line at individual genetic identity is a *subjective* choice, entirely dependent on a psychological frame of reference developed through the course of one's lifetime.

In the end, the purpose of this essay is to raise consciousness, with the hope of catalyzing radical change. To that end all we can do is provide food for thought. We hope that readers will consider that the existential crisis faced by contemporary civilization is caused by habitual cognitive entrenchment, as much as anything else. In developing the capacity for creating technological wonders, civilized humanity forgot how to *relate* to the world. It is time that we relearn what our animal ancestors knew instinctively, an innate intuition that our children still maintain up to a certain age—viz., that the universe is alive.

In times such as these continued evolution requires *revolution*.

# Chapter 1

# SYSTEM EARTH IN THE ANTHROPOCENE

This essay is about our relationship with the world, as seen in the context of evolution and human history. Through our accelerating acquisition and use of technological knowledge, our species has become a singularly powerful force of nature. Unfortunately, our ability to deal responsibly with that fact, and thus maintain a healthy relationship with the world that sustains us, has not kept pace.

We contend that Western civilization, in developing a global consumer economy based on industrial mechanization requiring rapid dissipation of non-renewable, high-grade energy, lost touch with reality and embarked on a path of self-destruction. Accessing a new path conducive to long-term human survival and quality of life will require that we fundamentally change our relationship with nature, which will in turn require that we significantly improve our comprehension of nature—including human nature. It will require that we develop a more realistic way of life, and healthier ways of imbuing our existence with meaning.

We are not alone in calling attention to the urgency of our situation. We do however have a unique explanation for how we got here, and the role of human intellect in that process. Contrary to what is now almost universally accepted as given, our technological creativity and scientific inquisitiveness have not served us well. The reason for this is that the development of our cognitive abilities produced an unhealthy mental imbalance. The technological aspect of the human mind has come to repressively dominate other aspects, and this is intimately linked to the unconstrained development of the consumer economy. Science and technology feed that system by supplying a continuous stream of 'disposable' commodities, as well as techniques for ensuring that people keep buying them, in order to drive economic growth, which then feeds back to drive science and technology. What many (perhaps most) people fail to appreciate is that this is a vicious cycle whose continuance assures the collapse of civilization, and quite possibly the extinction of humanity.

As will become clear in what follows, meaning is constructed by way of interpretation, and interpretation is a subjective matter. Depending on your perspective, the world can appear either simple or complex—and either very big or very small.

For instance:

Just before midnight on April 14, 1912 the "unsinkable" RMS Titanic, on her maiden voyage across the Atlantic, struck an iceberg and began to sink. The nearest land was 400 miles, and the nearest ship to receive the distress call 58 miles, about 4 hours, away. There were not nearly enough lifeboats for the approximately 2,200 people on board. During the two hours and forty minutes it took Titanic to sink, for those passengers—some 1,500 of who died that night—the world was very big.

On Christmas Eve 1968, Apollo 8 astronaut Bill Anders, orbiting the moon, took a picture of earth over the horizon. Three weeks later Anders' crewmate Frank Borman told the press, "I think the one overwhelming emotion that we had was when we saw the earth rising in the distance over the lunar landscape.... It makes us realize that we all do exist on one small globe. For from 230,000 miles away it really is a small planet."

Less than two years later, the third member of the Apollo 8 crew, Jim Lovell, was returning to the moon as commander of Apollo 13. Two days and 200,000 miles out an oxygen tank exploded and severely damaged the command module, rendering it inoperable. Fortunately, Lovell and his crewmates John Swigert and Fred Haise had a lifeboat, although it wasn't *intended* as such: the lunar module. In what ranks as one of the most remarkable feats of improvised ingenuity in human history, the three astronauts, with the aid of a team of NASA engineers on earth, were able to slingshot around the moon and successfully traverse the 230,000 miles to the one and only place that would allow them to continue living for more than a few more days. For the crew of Apollo 13, planet earth appeared small but loomed large.

A common thread linking these different stories and perspectives is the engineering of mechanical life-support systems enabling transport of humans across expanses of inhospitable space. Titanic and Apollo 13 were both marvels of human engineering designed specifically for that purpose.

In both cases however unanticipated disruptions of the mechanisms produced life-threatening situations. The disruptions were unanticipated because the mechanisms were built using models that were oversimplifications of reality. As we shall see, this will always be the case. Human models, and the mechanisms they engender, are simple, whereas nature is complex. And yet our way of life—a life-support system on which we now depend—is built on the assumption that our models adequately represent nature. As a result we are repeatedly surprised, and often alarmed, when it turns out that they do not.

And this is not all. We are living under the grand illusion that technological knowledge is all that is needed to wisely control events. This worldview is based on a web of misconceptions that stand or fall as one. We will show that it is replete with cognitive traps, and based on a foundation of questionable assumptions that prop each other up like a house of cards.

The following brief discussion of the development and evolution of life on earth is intended to bring into focus differences between the *mechanical* world that we have engineered and the *living* world that produced us, how the former has affected the latter, and our relationship to both.

The going will not be easy, for you will be asked to consider a point of view contra that which dominates the modern world, a worldview so engrained as to be considered common sense. And while the latter is a product of Western culture, its domain of influence is now global. The smallness of our world is very evident in the way we are rapidly becoming the "Global Village" that Marshall McLuhan wrote about fifty years ago (McLuhan, 1962). Planet earth—which includes the global human economy—is a *system*, and although it is easy to adopt the common practice of distinguishing and analyzing specific *parts* of that system in order to engineer solutions to our problems, vital information is lost in that reductionist approach. Central to our thesis is the contention that analysis of a complex system in terms of its component parts discards the very information required for understanding the system itself.

Earth's biosphere is a complex system on which we all depend for life and health, entailing relationships and interdependencies that are poorly understood, and which cannot be explained by appeal to isolated mechanisms. The converse is also to some extent true, in that earth's biology depends on the choices we make.

The impact of human activity on the biosphere cannot be understated. In a geological instant our species has gone from being a minor player in the biosphere to a major aspect of its functioning. We have at best a simple model of what humans have done and are doing to the complex interactions and feedback loops that constitute the eco-systems we are connected with. We are engaged in a massive global experiment, and while some of the results of this experiment are becoming more obvious every day, many remain largely unknown.

For most of earth's history that was not the case—earth came to life long before *Homo sapiens* arrived on the scene. And it will live on, in one way or another, even if we do not. Historically, extinction is the norm rather than the exception. We are nevertheless exceptional in that, for a single species, we have had an unprecedented impact. The by-products of our industrial technology have altered the climate and been recorded for posterity in the planet's geochemistry. This fact led earth scientists Paul Crutzen and Eugene Stoermer to propose that the contemporary geological era, termed the Holocene, be re-named the Anthropocene (Crutzen & Stoermer, 2000).

Earth has not always had a biosphere. Life as we know it emerged from a pre-biotic geochemical metabolism, which developed under the reducing atmosphere that enveloped earth some 4 billion years ago. However cellular life came into existence (and exactly how it did remains an unsolved mystery), once it did it began transforming the world toward its own ends. The transformation was not smooth, but rather involved a series of life-altering catastrophes. Of particular note was the crisis that followed the evolutionary 'invention', in cyanobacteria, of photosynthesis, which eventually produced an oxygenated atmosphere poisonous to most life at the time. One of the biggest mass extinctions in earth's history ensued, followed by an even bigger explosion of biological complexity. The latter was in no small part facilitated by endo-symbioses (Margulis, 1981), including those that produced the ancestors of mitochondria and chloroplasts—the sub-cellular organelles that provide the energy supporting all macroscopic plant and animal life.

The survivors of ecological catastrophes are those life forms that are able, through the development of cooperative relations, to take advantage of new opportunities (i.e., *potential*) that the crisis affords. In the case of the Great Oxygenation Event—perhaps the largest global environmental crisis in earth's history—the opportunity was a surfeit of free energy, which was exploited through the development of eukaryotic cells, and later, multicellularity.

Mass extinctions have occurred repeatedly over the past billion years, and each has been followed by a diversifying resurgence of life. There is no reason to believe that the extinction event occurring now as a result of human activities will be any different. What is in question is not the survival of the world, but of humanity.

Will contemporary civilization survive the Anthropocene? That is far from assured. There are several trends working against us: toxic pollution, climate change, decreasing species diversity, dwindling resources, escalating culture wars, and accumulating weapons of mass destruction. The most problematic trend, upon which all the others depend, is the explosion of the human population and concomitant development of an industrialized global consumer economy that celebrates *laissez-faire* capitalism. With this development we have insulated ourselves from nature, and in the process, lost touch with reality. Ironically, this can be directly attributed to the one human enterprise whose ostensive purpose is to distinguish reality from fantasy: *science.*

The problem is that some of the foundational assumptions of science are incompatible with the actual complexity of the real world, and fail to acknowledge its subjective dimension (Cilliers, 1998; Sagan, 2007). Because scientific models are interpretations that simplify through objectification, they at best describe only the externally differentiated *aspect* of nature (and an abstract *idealization*), which provides a grossly inadequate representation of life itself. So as science races ahead, it leaves the natural world further and further behind, creating in its stead an artificial, insular world of mechanisms. And that is the world in which we now live, and on which we now depend.

To be sure, science has provided deep insight into how the world works. The fruits of science—knowledge and technology—have worked wonders, and these wonders are what allowed the human population to explode. What we now take for granted would appear as magic to our ancestors. The problem is not that science is inherently bad; it is that it is inherently *limited*. Its limits have not been widely or adequately acknowledged. As a result of this *hubris*, humanity, like Dr. Frankenstein, has lost all perspective. The common-sense notion that health entails balance is lost on the industrialized world.

Like the biosphere on which it depends, the Global Economy is a complex system. But unlike the biosphere, which runs on the virtually inexhaustible but relatively diffuse energy of the sun, the Global Economy owes its existence to the dissipation of concentrated deposits of non-renewable, high-grade energy—petrochemical and nuclear fuels. In developing this economy, the human population grew dependent on its short-term (and quite temporary) productivity, which depends in turn on industrial dissipation. As that dissipation produces toxic waste that poisons the water we drink, the food we eat, and the air we breathe, and as extraction of the requisite fuels entails wholesale destruction of land, water, and wildlife, the Global Economy is rapidly destroying the biosphere. And yet, changing course—our way of life—is far easier said than done. As has oft been noted, the problem is essentially one of *addiction*, except on a global scale. The dependencies entrained by the Global Economy have developed lives of their own: we do not control them; they control us (Hayles, 1999; Arthur, 2009).

Over the past two years two monumental disasters brought our predicament into sharp focus. The first was the explosion of the Deepwater Horizon oil extraction platform in the Gulf of Mexico, producing a vast and deadly oil spill that became the worst environmental disaster in American History. The second was the earthquake and tsunami in Japan that ruptured containment at the Fukushima Daiichi nuclear power plant, releasing huge amounts of radioactive waste into the ocean and atmosphere. These two events, as well as others like them in recent decades, inflicted untold harm on the biosphere, the full extent of which will never be known.

Despite these catastrophes, offshore drilling and development continue around the world, as does the implementation of nuclear power. There is, as far as we know, simply no other way that the Global Economy can sustain itself. And if the Global Economy fails to sustain itself, a significant proportion of the human population will no longer have any means of support, and as a consequence, will suffer. Widespread famine and escalating wars will likely occur. But significant change of any kind entails some amount of suffering, for somebody somewhere. And therein lies the problem: as much as we may want things to change, no one wants to suffer, even a little bit for a little while. This includes the 'movers and shakers' who are comfortably ensconced in wealth and the lion's share of political power. So what ought to be done today gets put off until tomorrow.

Unfortunately, the reality is that *significant change will happen whether we like it or not*. The question then is not how to prevent suffering, but rather, how to minimize it. We need to be thinking about lifeboats, carrying capacity, and what kind of world we want for our children and future generations. We need to realize that the world we have created is not grounded in reality—that the industrialized global consumer economy amounts to Global Insanity.

To see this all one need do is pause for a moment to think about the amount of trash produced by the average American household. How much do you generate each week? Now multiply that 300 million times. Where does it all go? What happens to the toxic waste from plastics, batteries, electronics, etc.? Chances are either you don't know, don't care, or simply don't want to think about it. But is that acceptable? Is denial of reality a sane way to live?

Global insanity has many manifestations, epitomized by the culture and technology of war. A common thread running through all of them is the absence of *care*. We say we care, but for most of us that is lip-service. The stark reality is that the Global Economy in which most of us are active participants is violent, self-serving, and ultimately self-destructive. It is run by multi-national corporations that lack the capacity to care: shareholders generally do not pay for, much less experience, most of the real costs of economic activities, which are mere 'externalities'.

Civilization is in dire straights, and denial is a natural psychological defense against despair. The denial is widespread and takes many forms (Magdoff & Foster, 2011). In the United States some forty percent of the population continues to deny the fact of evolution. A similarly large percentage denies the fact that human activity is changing the climate. Such ignorance is political putty in the hands of the plutocrats who own most of the world. Meanwhile, the scientific enterprise spends billions of dollars on unsustainable, environmentally toxic research and development that is of practical benefit only to those economically privileged enough to afford the short-term, high-tech fixes that such work generates. And yet, this enterprise is sold as the savior of humanity.

Such delusions understandably lend comfort, but they are maladaptive and will only exacerbate the suffering that looms on the horizon. The Global Economy has seriously damaged our life-support system (Eagleton, 2011). Are we thus doomed, like the Titanic? Or can we turn things around, like Apollo 13? Prospects for the latter appear exceedingly low, as it would require radical psychological, social, and cultural change based on a complete re-evaluation of cherished beliefs and deeply held assumptions. It would require that we acknowledge and accept that our models do not and cannot adequately represent the real world.

As of this writing in the summer of 2012, the U.S. is experiencing the worst drought in over 50 years, and the polar ice caps and mountain glaciers are melting at a record rate. There is little doubt that these events are linked to anthropogenic climate change, which is also causing an increase in the magnitude and/or frequency of extreme weather. When we began writing in the spring of 2011 the Southern US was ravaged by tornados. Such natural disasters remind us that survival is never assured. After all, the dinosaurs were wiped out when an asteroid collided with earth, and the possibility exists that the earth will at any moment be fried to a crisp by a gamma ray burst from binary star 8,000 light years away. But where there is life there is uncertainty; so it is possible, if not probable, that *Homo sapiens* will survive. Which raises another, perhaps more important question that we should all be asking ourselves: if we do survive what will be the quality of life?

We submit that if civilization is to survive the Anthropocene—and if it is to enhance rather than degrade the quality of human life—humanity needs to grow up, and fast.

Paradoxically, the only way that that can happen is for civilization to become *less* mature. We need to reclaim our lost youth, which requires some deconstruction.

# Chapter 2

# OF METAPHORS, METAPHYSICS, AND MATH: A MYTHOLOGY OF MECHANISMS

Insight into our predicament requires an understanding of knowledge itself: how it is acquired, and how, by shaping our values, it influences our relationship with the world. To address this we need to consider how we use language to encode, interpret and communicate our perceptions. We can then ask how this linguistic way of knowing informed the historical trajectory of thought that gave rise to Reductionism, the widely held Western belief that the world can be adequately understood in terms of mechanisms elucidated by taking things apart.

Human beings are social creatures, and our unique consciousness is socially developed. Underpinning this development is mythology (Campbell, 1991), a widely shared collection of often literally-interpreted stories that metaphorically describe human experience, establishing a framework for enculturation. The psychological development through which we each become cognizant of ourselves as mortal beings separate from the world, and of how we are supposed to deal with that uncomfortable fact of existence, is linguistically entrained via mythological narrative. In other words, we learn to think about ourselves and our relationship with the world through the stories we are told when we are growing up. Those cognitions are reinforced throughout our lives by the stories we tell ourselves and each other, stories that metaphorically resonate with those of our childhood.

George Lakoff and Mark Johnson (2003, 1999) have discussed the fundamental centrality of metaphor—the use of one thing to describe another, completely different thing—as a linguistic device for communicating experience and expressing self-awareness. It is hard to imagine how we could think or communicate about anything without using metaphors. To see this consider what you just read: to 'imagine' anything is to form an 'image'. But not a *literal* image; it is an *analogue*: we metaphorically 'see' things in our mind's 'eye'. But what we are describing here is completely different from *literally* seeing with our *actual* eyes.

If you are finding this difficult to 'grasp', you might say that it is 'beyond me', 'over my head', 'too deep' or 'hard to fathom'.

At some point in your life you may have had your 'heart broken'.

A hunch is a 'gut feeling'.

And so on. You get 'the picture'.

Metaphorical language is one of the two fundamental ways we encode our models of reality, and the primary way that we both conceive and communicate complex perceptions. In fact, many words with meanings that we now take to be literal began as metaphors. For example, the word 'understand' originally meant to stand in the midst of (with *under* having once had a 'broader' meaning than it does now), and its synonym 'comprehend' originally meant to completely catch hold of (*com-*, from complete, and *-prehend*, from the Latin prehendere, as in 'prehensile').

Most of our metaphor-based cognitions are not even conscious. When we make a decision, we may think we are consciously thinking things through, but more often than not what we are doing is *rationalizing* (that is, reflectively selecting bits and pieces of knowledge to come up with a reason for) a choice that we already made subconsciously. These subconscious decisions are themselves rooted in the mythologies informing the moral 'compass' through which we interpret experience. Lakoff (2008) has written about how this guides our decision-making process in politics—and more importantly, how it is exploited by savvy politicians and their backers (not to mention preachers and other influential public speakers). The idea, called 'framing', is that value-laden interpretations of otherwise neutral facts are subconsciously constructed when those facts are 'framed' by metaphorically evocative language.

The problem, like so many others explored here, is developmental: the cognitive frames that each of us is endowed with are constructed during and by way of our psychological development. Conscious comprehension of anything requires the development of an appropriate cognitive frame, which depends, in part, on our upbringing—our level of education, and more importantly, what stories we were told, paid attention to, and 'resonated' with. This explains why otherwise intelligent people will often deny empirically demonstrable facts (such as evolution, which has been *framed* as an atheistic worldview by outspoken theists and atheists alike), and can be easily induced to vote against their best interests. It also explains why advertising is so effective, not to mention lucrative.

Given that consciousness as we know it developed in concert with language and is intimately linked to the invention of metaphor, it is worth considering how, in the course of human evolution, this unique human faculty came to be. A theory that bears consideration is that of the late Julian Jaynes. In his remarkable book *The Origin of Consciousness in the Breakdown of the Bicameral Mind*, Jaynes (1976) proposed that metaphor-based self-referential cognition emerged quite recently in human evolution—only ~3,000 years ago, about the time that Homer's *Odyssey* was written. Prior to that, as late even as the events recorded in the *Iliad*, human beings were un(self)consciously motivated, either by habit (in routine circumstances), or (in novel or stressful situations requiring decisions) by poetic directives generated in the right hemisphere of the brain that were received in the auditory center of the left hemisphere, and thus literally *heard* as voices, which were interpreted as coming from an external authority, i.e., a god. These authoritarian auditory hallucinations were perhaps not unlike those experienced by modern-day schizophrenics, and in this light the latter (as well as the modern susceptibility to hypnotism) might be viewed as an atavism. The major difference between then and now was that attending to hallucinated 'voices' was then the cultural norm, and hence not considered to be insane.

According to Jaynes, the evolutionary 'breakdown of the bicameral mind' and concomitant emergence of the metaphor-based consciousness was catalyzed by the collapse of ancient civilizations—whose hierarchical sociocultural structures had developed around (and reflexively reinforced) hallucinated auditory authorization—in the face of anthropogenic ecological crises, which caused widespread famine and hence cultural conflict attendant on mixing of displaced populations attending to different 'gods'. The transition is chronicled in ancient texts, including those of the Old Testament, which document the progressive silencing of the divine voices and refinement, through writing, of metaphorical narrative as a new device for modeling experience and articulating a code of morality.

Thus the Judeo-Christian myth of the fall from grace, precipitated by Adam and Eve eating the fruit of the Tree of Knowledge, is for Jaynes a metaphor for the emergence of metaphor-based (self)-consciousness, which resulted in a loss of innocence. Consciousness rationalizes conscience, or at least the value-laden sense of personal responsibility that informs our actions. But how the latter developed clearly varied between cultures. Native American scholar Vine Deloria Jr. (2003) has pointed out that in the Christian mythology adopted by Western Europeans, the fall affected not just humanity, but all of nature. In this myth humans are left with the thankless task of having to overcome and manage a debased (and shameful) nature. Eco-philosopher Freya Mathews (2003) argues that the cognitive strategy that develops to that end is *repression*, which is facilitated by Cartesian dualism (which we will take up below). The Native Americans (and other aboriginal or 'pagan' cultures) developed a different myth, in which human beings are viewed as part of (indeed owe their existence to) nature, which is celebrated. Thus, our current dysfunctional relationship with the natural world can be seen to extend from certain religious metaphors that came to repressively inform Western consciousness.

Although it remains controversial and is in some ways problematic, Jaynes's theory merits serious consideration given the well-documented functional differentiation of the brain hemispheres and the undeniable role that metaphorical language has played in the evolution of human cognition. Moreover, it offers a compelling explanation of many otherwise mysterious psychological and historical facts. But for present purposes it serves only as backdrop for three rather uncontroversial postulates: first, that human experience is modeled ('molded') and communicated, and our collective knowledge thus produced, through metaphorical narrative; second, that this was not always true, because the defining characteristics of humanity—including the characteristic way that we use metaphor to create knowledge—must have developed (like everything else) from ancestral precursors that *lacked* those same characteristics; and third, that what is considered normal in one cultural context might well be considered *insane* in another, perhaps more evolved context. We will consider each of these postulates in due course, starting with the last, which echoes the title of this book.

It is not a stretch to imagine that what we accept as normal today will be considered insane by our descendents. In fact, it is not a stretch to say that already, to many of us, the socio-cultural norms of Western civilization appear completely insane. And this may not be 'mere' metaphor—it may to some extent be literally true. In his recent book *The Master and his Emissary: The Divided Brain and the Making of the Western World*, psychiatrist Iain McGilchrist convincingly argues that modern reality both stems from and reflects an imbalance between the brain hemispheres that developed over the past few hundred years (and especially since the Industrial Revolution), which has many striking similarities to schizophrenia, a mental condition associated with right hemisphere deficiencies. Thus, while McGilchrist agrees with Jaynes that the evolution of human consciousness involved changes in the relationship between the two hemispheres of the brain, he holds that Jaynes got it backwards: that with language-based consciousness the mind became progressively more divided, rather than less so. Our ancestors may well have hallucinated divine voices, but this was not a schizophrenic condition—schizophrenia is, as McGilchrist notes, a relatively modern malady that has dramatically (and tellingly) increased in prevalence over the past two hundred years.

At this juncture it is worth noting that the initial draft of this essay (see Coffman & Mikulecky, 2012, for a published excerpt) was written prior to our having read or even been aware of McGilchrist's book. Since his thesis resonates strongly with ours and is especially pertinent to the discussion of this chapter, we will spend some time reviewing it here.

Like Jaynes, McGilchrist seeks to understand how the well-known structural and functional differentiation of the two brain hemispheres (left and right) relates to human experience. The basic idea, supported by extensive scientific literature on brain function in humans and other animals (which also have hemispheric differentiation), is that each hemisphere attends to the world in fundamentally different ways. The right hemisphere is widely vigilant—attending to the 'big picture', and seeking out the 'other'. In contrast, the left hemisphere is more narrowly focused—attending to immediate technical problems and the 'task at hand'. This is true in other vertebrate animals (including e.g., birds) as well as in humans. In humans however the focused left hemisphere has among other things come to specialize in language, which provided it with a remarkably useful tool for accomplishing its job of 'manipulating' the world. In the meantime the right hemisphere remains more 'grounded' in the immediate experience of the actual (phenomenal) world, and according to McGilchrist, is required to bring the real world 'into being' for the left hemisphere. This is supported (for example) by the fact that patients or subjects with left hemisphere deficits have no problem drawing relatively realistic pictures of what they see, whereas those with right hemisphere deficits tend to draw pictures that are disjointed or partial. Moreover, the latter subjects (and many schizophrenics), unlike the former, often experience parts of their own body as being alien, to the point even of claiming that they belong to someone else and not themselves (McGilchrist, 2009).

So although normally the brain as a whole is functionally integrated, it, like the hands it controls, manifests left-right differences in abilities and preferences (as we will see in Chapter 4, such symmetry-breaking differentiation is the normal, expected outcome of any system-level development). As a consequence of this hemispheric differentiation, our grounding in reality is largely dependent on the right hemisphere ('The Master'), whereas our ability to interact productively with and manipulate the world is largely dependent on the left hemisphere ('The Emissary'). In line with this, the right hemisphere takes a 'gestalt' or 'holistic' view, and interestingly enough, tends toward melancholy and pessimism (what some of us refer to as 'realism'), whereas the left hemisphere takes a more fragmented, focused, and reductionist view (which indeed it needs to do in order to 'get the job done'), and tends toward mania and optimism (which some of us refer to as 'rose colored glasses'). Concomitantly, the right hemisphere is comfortable with fluid vagueness, that which is implicit and uncertain—and hence, with the intuitive and metaphorical—whereas the left hemisphere prefers crisp fixedness, that which is explicit and certain—and hence the analytical and literal. It should be obvious that health requires that the two work together in a well-balanced relationship.

But unfortunately that does not always happen, and in fact may be increasingly difficult to achieve, given that the right hemisphere depends on the left hemisphere to engage with the world. With human beings that entails the use of language, a left hemisphere specialty. The problem is that the use of language to model reality creates the illusion that the developed model is capable of fully representing reality; as a result, the 'Emissary' (the left hemisphere) comes to think it knows best, and thus to repressively dominate, the 'Master' (the right hemisphere). In McGilchrist's words:

*The existence of a system of thought dependent on language automatically devalues whatever cannot be expressed in language; the process of reasoning discounts whatever cannot be reached by reasoning. In everyday life we may be willing to accept the existence of a reality beyond language or rationality, but we do so because our mind as a whole can intuit that aspects of our experience lie beyond either of these closed systems. But* in its own terms *there is no way that language can break out of the world language creates—except by allowing language to go beyond itself in poetry; just as* in its own *terms rationality cannot break out of rationality, to an awareness of the necessity of something else, something other than itself, to underwrite its existence—except by following Gödel's logic to its conclusion. Language in itself (to this extent the post-modern position is correct) can only refer to itself, and reason can only elaborate, 'unpack' the premises from which reason must begin, or that validates the process of reasoning itself—these premises, and the leap of faith in favour of reason, have to come from behind and beyond, from intuition or experience.*

*Once the system is set up it operates like a hall of mirrors in which we are reflexively imprisoned. Leaps of faith from now on are strictly out of bounds. Yet it is only whatever can 'leap' beyond the world of language and reason that can break out of the imprisoning hall of mirrors and reconnect us with the lived world.* (pp. 229-230; emphasis in original)

Alicia Juarrero (1993) has explored how this played out historically beginning in Greece, as mythology (which according to Jaynes originated poetically in the right hemisphere) gave way, via progressive de-contextualization, to philosophy (which according to McGilchrist is a product of the analytical left hemisphere) as the preferred way of explaining the world. But by its own logic philosophy ultimately produced the recognition that language cannot possibly refer directly to any reality but its own. This of course led directly to postmodernism.

But the modern world is a product of left hemisphere successes. So postmodernism, beyond being viewed as a political nuisance emanating from the ivory tower of academia, is given little credence (as would be expected given the left hemisphere's talent for focusing attention by repressively ignoring things it deems irrelevant to the utilitarian 'task at hand'). In the modern worldview, which is still very much the dominant worldview, language is generally considered to provide a transparent window on an 'external', objective reality, of which we are mere 'observers'. The idea that, through language, we are actually active participants in the creation of reality (Juarrero, 1993) still seems lost on the majority. And according to McGilchrist, this is to be expected given the predilections of an increasingly dominant left hemisphere:

> *So if I am right, that the story of the Western world is one of increasing left-hemisphere domination, we would not expect insight to be the key note. Instead we would expect a sort of insouciant optimism, the sleepwalker whistling a happy tune as he ambles towards the abyss.* (p. 237)

In other words, since the right hemisphere is responsible for keeping us grounded in reality, repression of that hemisphere by the left, whose written language-mediated ascendancy began in antiquity (resulting in 'silencing' of the poetically 'divine' voices of the right hemisphere, as proposed by Jaynes), continued over the past few millennia, and was accelerated by the Industrial Revolution, would be expected to result in human beings progressively losing touch with reality—much as in fact happens in the extreme with schizophrenia. 'Global insanity' may thus be more than just metaphorically descriptive of our current situation—it may well have some literal truth.

So we can summarize and extend what we have developed so far as follows: first, all human knowledge is both based on and *framed in* metaphorically evocative language developed by way of mythological narrative, and is thus unavoidably *value-laden*; second, knowledge is creatively realized within the world, in a participatory fashion, and hence is more or less *realistic*; third, the realization of knowledge has long-term as well as short-term consequences, some intended, others not; and fourth, this process necessarily involves two differentiated brain hemispheres that attend (and thus relate) to the world in fundamentally different ways: one (the right hemisphere) doing so widely, intuitively (i.e., with immediacy), and metaphorically, the other (the left hemisphere) doing so narrowly, analytically (i.e., at a disengaged 'distance'), and literally. Finally, there is evidence that the left hemisphere, via its facility with language and by way of socio-cultural feedback, has come over the past few centuries to repressively dominate the right hemisphere (McGilchrist, 2009).

Our use of words therefore matters greatly:

> *The words we use to describe human processes are highly influential for the way we conceive ourselves, and therefore for our actions and, above all, for the values to which we hold. With a rising interest in neuroscience, we have an opportunity, which*

*we must not squander, to sophisticate our understanding of ourselves, but we can only do so if we first sophisticate the language we use, since many current users of that language adopt it so naturally that they are not even aware of how it blinds them to the very possibility that they might be dealing with anything other than a machine.* (McGilchrist, 2009: 459)

In the second half of this chapter we discuss how that blindered, increasingly unbalanced worldview developed historically, ultimately giving rise to the modern world. In the end, the question of whether the patent insanity of that world is literal or metaphorical is not all that important, and quite beside the point. For, as McGilchrist concludes:

*The divided nature of our reality has been a consistent observation since humanity has been sufficiently self-conscious to reflect on it.... When one puts that together with the fact that the brain is divided into two relatively independent chunks which just happen broadly to mirror the very dichotomies that are being pointed to—alienation versus engagement, abstraction versus incarnation, the categorical versus the unique, the general versus the particular, the part versus the whole, and so on—it seems like a metaphor that might have some literal truth. But if it turns out to be 'just' a metpahor, I will be content. I have a high regard for metaphor. It is how we come to understand the world.* (pp. 461-462)

The other way we come to understand the world (that is, to consciously model reality) is by using formal logic, as epitomized by math (Lakoff & Nunez, 2000). Math is itself a language, but it is unique in its affordance of a *precise* way to represent and encode abstract ideas describing the physical dimensions and quantitative properties of the world, and to subject those ideas to rigorous tests for logical consistency. It is a formal means of constructing chains of entailment. Math thus counterbalances with crisp precision the vagueness of metaphorical language (which is just as effective, through fiction and rhetoric, at creating fantasy or reinforcing delusion as it is at modeling reality). Some might even say that math (or more generally, formal logic) provides the universal test of Truth: the one means we have of determining, with absolute certainty, whether imagined ideas about reality are true or false.

While this may (or may not) be true, it is true that some truths are axiomatic (true by definition, e.g., 1+1=2), which allows other less obvious truths to be *proved* mathematically. But Kurt Gödel proved mathematically that some truths cannot be proved mathematically. That is, no formal system (such as math, or more generally language) that is *consistent* (meaning that it can not be used to prove that the same statement is both true and false) can also be *complete* (able to prove all its postulates and theorems). So, either some truths lay forever beyond our ability to *know* with certainty, or math (and for that matter language) is not the only route to knowing truth.

As discussed by Rosen (2000), math is intimately tied to measurement, the quantitative evaluation of qualitatively definable things, either in relation to one another or to some standard. For example, through measurement the relationships manifested by geometric forms can be expressed as algebraic equations, and thence manipulated using arithmetic. A classic example is the Pythagorean theorem relating the length of the hypotenuse ($h$) of a right triangle to the length of the sides ($s_1$ and $s_2$): $h^2 = s_1^2 + s_2^2$. Pythagoras, struck by the fact that vibrating strings on musical instruments produce harmonies when their lengths are commensurably shortened (e.g., by the ratio 3:2, producing the dominant fifth), came to believe that any quality (such as a musical pitch) can be quantified, i.e., represented as a number. This in turn led to the belief that everything is measurable, and conversely, constructible from measured units (Rosen, 2000).

This belief, which is widespread to this day and is one of the foundations of Reductionism, sprang from an initial assumption: that any two line segments are commensurable, that is, have lengths whose ratios are a rational number. While this assumption appeared reasonable at the outset, it was upended by facts that manifested when the geometric space was enlarged from one dimension to two: for example, the Pythagorean Theorem shows that $h/s$ (the square root of 2) is not rational. The Pythagorean solution to this problem was to enlarge mathematics to include irrational numbers, which assign values to things that are not commensurable and hence not countable. The problem with this however, as shown by Zeno's paradoxes, is that introducing incommensurability breaks the tidy reciprocal relationship between measurement and construction, by extending mathematics to the immeasurable limit of infinity. Unfortunately, we seem still to not have grasped the implications of this. Rosen (2000) sums up the situation as follows:

*As we have seen, commensurability was the original peg, the* harmonia, *that tied geometry to arithmetic. It said that we could express a geometric quality, such as the length of an extended line segment, by a number. Moreover, this number was obtained by repetition of a rote process, basically counting. Finitely many repetitions of this rote process would exhaust the quality. But more than this, it asserted that we could replace the original geometric referents of these numbers by things generated from the numbers themselves, and hence we could completely dispense with these referents....*

*The repetition of rote operations is the essence of algorithm. The effect of commensurability was to assert an all-sufficiency of algorithm. In such an algorithmic universe, as we have seen, we could always equate quality with quantity, and construction with computation, and effectiveness with computability.*

*Once inside such a universe, however, we cannot get out again, because all the original external referents have presumably been pulled inside with us. The thesis in effect assures us that we never need to get outside again, that all referents have indeed been internalized in a purely syntactic form.*

*But commensurability was false. Therefore the Pythagorean formalizations did not in fact pull qualities inside. The successive attempts to maintain the nice consequences of commensurability, in the face of the falsity of commensurability itself, have led to ever-escalating troubles in the very foundations of mathematics, troubles that are nowhere near their end at the present time.*

In the twentieth century several mathematicians, who included Alan Turing and Alonzo Church, worked to formalize the idea that anything that is algorithmically computable can in fact be computed by a machine (i.e., a 'Turing Machine', now known as a computer). Although Turing himself recognized that not everything is computable, implicit in Reductionism is the notion that it is. Rosen (2000) continues:

*What we today call Church's Thesis began as an attempt to internalize, or formalize, the notion of effectiveness. It proceeded by equating effectiveness with what could be done by iterating rote processes that were already inside—i.e., with algorithms based entirely on syntax. That is exactly what computability means. But it entails commensurability. Therefore it too is false.*

*This is, in fact, one way to interpret the Gödel Incompleteness Theorem. It shows the inadequacy of repetitions of rote processes in general. In particular, it shows the inadequacy of the rote metaprocess of adding more rote processes to what is already inside.* (pp. 77-78; emphasis in original)

If you are paying attention you no doubt noticed a remarkable congruence between this statement of Rosen and that of McGilchrist quoted above, regarding the imprisoning limitations of language-dependent systems of thought (which include math). The upshot is that the real world, which manifests an infinity of 'aspects' within any given locale, is always greater than the representational capacity of any syntactically formal system that arises within it, and is therefore not computable. Computation can only be carried out using algorithms repeated a finite number of times, which can at best simulate, and never re-create, the world. And yet, the "unreasonable effectiveness of mathematics in the natural sciences" (Wigner, 1960) seduces us into thinking otherwise. Moreover, it allows us to create a *virtual* reality that keeps us endlessly entertained by illusion, as epitomized by modern computationally-augmented movies, games, and other media.

Be that as it may, math certainly appears, by virtue of the insight it affords into the physical world and the wonders of engineering that it uniquely enables, to be the most powerful means we have at our disposal for elucidating certain universal truths, and its use has gotten us to where we now are. In the following we will briefly trace the historical trajectory of that development, through some of its major figures, in order to set the stage for the argument that follows: that the Western way of knowing and hence shaping reality—the way of contemporary *science*, upon which our Global Economy is founded—is seriously misleading.

### Plato and Aristotle

The ascendancy of Western science began in Greece. Thanks to the curatorial scholarship of Christian theologians during the 'dark ages' following the fall of Rome, which made possible the revival of classical culture during the European Renaissance, we in the Western world inherited our secular intellect from the ancient Greek philosophers. Although there were competing schools of thought promoting different worldviews (e.g., that of Heraclitus, whose ideas concerning perpetual change have enjoyed a recent renaissance and presage much of the discussion in this essay), we will focus here on two towering figures who ultimately came to dominate Western discourse: Plato and his student Aristotle. Through the metaphorically grounded *metaphysics* articulated in their extensive writings, both played a major role in creating the world as we know it.

The metaphysics of Plato (428-348 BC) grapple with the nature of reality. The metaphysical dualism that polarizes philosophy to this day echoes Plato's conception that the concrete world of matter and the abstract world of ideas constitute two separate realities. For Plato truth was itself an idea, as were common characteristics that allow us to categorize specific objects into generic classes. The objects themselves were then imperfect manifestations of what became later known as 'Platonic Ideals'. For example, a line drawn in the sand (or anywhere else) is an imperfect manifestation of the idea of (i.e., the ideal) Line. The famous metaphor offered in *The Republic*, one of many used by Plato to dialogically articulate his philosophy, was that of the Cave. On the walls of the Cave in which we are chained (metaphorically representing the dependency of experience on our senses) we see shadows of objects, cast from behind us by the light outside. The shadows are what we interpret as reality. But the shadows are just images projected on our senses, and the Reality that they represent is that of eternal, changeless forms—which include those that are elucidated mathematically—that lie outside of our Cave of perception.

With Plato we thus see a metaphysical rift between the ideal or 'internal' world of mind and the actual or 'external' world of objects, a schism that vexes human consciousness to this day. The Platonic formulation of this subject/object dualism, as articulated in the metaphor of the Cave, is the inverse of its modern formulation, in that the external world for Plato is composed of ideal forms whereas the internal world is that of particular objects. Since for Plato the former are more real than the latter, in his metaphysics we find an early expression of *idealism* or *rationalism*.

Like Plato, Aristotle (384-322 BC) was interested in elucidating universal truths or principles by which the workings of the world can be rationalized. But unlike his mentor, Aristotle believed that such principles exist within the *nature* of things and can thus be discovered by studying their particular instantiations, as opposed to their ideal forms. He thus championed 'inductive' reasoning from observation of nature, in addition (and in contrast) to the 'deductive' reasoning from axiomatic first principles and formal logic (as in math) championed by his mentor. For Aristotle the world of the senses was real and instructive, and in his metaphysics we find the foundations of *realism* or *empiricism*, the basis for Western science.

Among other things, Aristotle sought to articulate logical accounts of observed phenomena—that is, to explain them by identifying their *causes*. A cause is simply something on which the occurrence of something else depends, or to which it can be realistically attributed—that which makes something happen. Aristotle recognized four distinct causal categories: material, efficient, formal, and final. Material cause is simply the *substance* of which something is made, on which its existence depends. Efficient cause is the *action* that that makes it happen. Formal cause is the set of *circumstances* that enable or entrain its occurrence. And final cause is its consequence or *purpose*, the need fulfilled by way of its occurrence. This is classically illustrated by a house: the material cause is the bricks, mortar, and other materials that are used in building; the efficient cause is the labor that goes in to putting those materials together; the formal cause is the design ('blueprint') that is being executed; and the final cause is (fulfillment of) the need for shelter. According to Aristotle all four categories are required to adequately explain any given phenomenon in nature. Aristotle further stipulated (problematically as we will see) that a thing cannot cause itself (Juarrero, 1999).

The mental abstraction of general categories or principles, such as Ideal Forms and Causation, from specific things and actions in the actual world—a dichotomy, created by metaphorical and mathematical encoding in human language, which forces the perennially vexatious questions about the nature of reality to which we shall turn in moment—was thus seminally expressed in the works of two of the principal architects of Western thought, Plato and Aristotle.

To begin to see how this linguistic way of knowing transformed the world we now turn to three influential thinkers of the European Renaissance and Enlightenment.

### Bacon, Descartes and Newton

Francis Bacon (1561-1628) sought to systematize all knowledge. Toward that end he articulated a rulebook for natural philosophy that ultimately became known as the Scientific Method. For Bacon, the stated purpose of natural philosophy was to increase knowledge, and thus power. Knowledge of *causes* does that, because if you know the cause of something, you can devise ways to predict its occurrence, and intentionally make it happen, prevent it from happening, or shape it to your own ends. Of the four Aristotelian causal categories, Bacon considered only three—material, formal, and most especially, efficient—to be relevant to this purpose, because together they suffice to reveal the objective nature of things, and more importantly, *how* they work. For Bacon, final cause, the *need* that something fulfills—the reason *why* it occurs—is strictly a subjective human concern and has no place in natural philosophy, because philosophy is not suited to gleaning God's (or anyone else's) subjective purposes. That is the job of theology. And to this day, final cause, and indeed the question *why*, is generally not considered to be a legitimate concern of science.

Bacon did not invent the Scientific Method. Others, including Alhazen, da Vinci, Copernicus, Kepler, and Galileo, probably deserve more credit than he for that. But he played a major role in codifying the scientific approach to knowledge production, emphasizing the central role of empirical verification through experimental testing. Although he rejected Aristotelian final cause, he embraced Aristotelian empiricism with its inductive reasoning. He also recognized that, when it is left unchecked, knowledge is biased by subjectively irrational beliefs and prejudices, which he referred to as Idols. His approach sought to systematically eliminate the influence of Idols, in order to obtain a more impartial or 'objective' view of the world.

And in doing so, through his influence as a scholar, statesman and writer, he contributed significantly to the ongoing human *objectification* of reality. For Bacon, practical knowledge of the natural world, gained through study of its particulars, is valuable because of the power it grants: the power to control, to bend the world toward one's own will. One might say that the purpose (final cause) of Bacon's system of knowledge is to serve the selfish need we each have to control the course of events. Toward that end, knowledge is most easily gained by manipulating and dissecting the world, which requires that we treat its inhabitants—including other creatures, and even other human beings—as objects.

Of course, bending the world toward one's own purposes is nothing new; nor is it by any stretch a uniquely human activity. It is a fact of life: all organisms do it, many of them violently. But only humans do it consciously and hence *rationally*. By removing final cause from consideration, Bacon's system of practical knowledge—a significant part of the foundation of the modern scientific industry—further differentiated subject and object, a process that had begun with the development, via metaphor, of human consciousness, and which became increasingly defined through metaphysical discourse. Through the metaphysics of Bacon, human beings became a bit less inclined to view the world and its inhabitants sympathetically as subjects in their own right, i.e., agents with ends of their own, and more inclined to view them as mere objects to be studied and manipulated for selfish purposes. Moreover, by rejecting final cause, Bacon devalued the consideration of consequences in scientific efforts to acquire knowledge. This contributed to the development of *ex post facto* rationalization as a cognitive approach to dealing with the world, making it easier to take without asking permission. Science became a means to dominate Nature, "to conquer and subdue her, to shake her to her foundations". In other words, science enabled rape of the earth.

René Descartes (1596-1650), a contemporary of Bacon whose rationalism (recalling Plato) complemented Bacon's empiricism (recalling Aristotle), further codified the schism between subjective *self* and objective *other* by declaring that mind (also known as soul, spirit, or in Greek, *psyche*) and matter (the body, or Greek *soma*) are fundamentally different things that happen to come together within the brain of a human being. Non-human animals are supposedly not so fortunate, being but soulless bodies (without mind). The material body is solid and spatially extensive, whereas the immaterial mind is neither. But if the mind is not made of matter, how do we know it exists? Descartes answers confidently: *cogito ergo sum*—I think, therefore I am. For Descartes this mental 'I' that obviously exists is not the body, it is the soul. Although it remains anyone's guess as to why body and soul must be separate things (as discussed by McGilchrist, 'disembodiment' is a predilection of the left hemisphere that is often experienced by schizophrenics), this infectious notion, which was first articulated and discussed at length by ancient Greek philosophers and according to Jaynes emerged concomitantly with the development of metaphor as a linguistic device for describing subjective experience (allowing 'I' to be objectively conceived for the first time), thus became cemented into Western consciousness as *Cartesian dualism*.

It is instructive to view this metaphysical development in socio-political context. The Roman Catholic Inquisition was alive and well in Descartes' time, and in 1633 tried Galileo for heresy. Descartes was no fool. His separation of body and soul served to placate the church with the assurance that science is only concerned with the body. So the church retained authority on the soul. This had much the same effect as Bacon's consignment of final cause to theology.

And thus was born an unconscious cognitive collusion between science and religion that has been instrumental in shaping our reality, for reasons that will become more apparent in the discussion that follows.

But a reified mind-body dualism is not all that Descartes left us: for in constructing a metaphysical framework that cleanly separates mind from matter he helped advance the emerging reductionist notion that the living body—and the material world in general—is a *machine*. Ironically (given their divergent metaphysical positions) this too fits Bacon's agenda like a glove, meshing so well that one might be forgiven for wondering whether that fact alone can be taken as evidence for the veracity of this schizoid approach to being in the world.

A final irony is found in the fact that modern science maintains Descartes' mechanistic view of life while rejecting his dualism, proclaiming instead that *mind* is an illusion and does not really exist at all—a claim that Descartes himself would rightly have rejected as patently absurd. For, to his credit (he was, after all, a brilliant philosopher), Descartes realized that the only possible route to conceiving life as a machine is via mind-body dualism. The former logically entails the latter.

Isaac Newton (1643-1727) sealed the deal with his masterpiece *Philosophiae Naturalis Principia Mathematica*: the mathematical principles of natural philosophy. Newton's three laws of motion, precisely articulated using the new calculus that Newton himself invented, allowed anyone with the requisite information and mathematical skill to predict the exact trajectories of interacting particles. The first law states that the velocity of a particle does not change unless the particle is acted on by an external force. The second law states that force equals mass times acceleration: $F = ma$, where acceleration 'a' is the temporal derivative of velocity, i.e., its rate of change with time. The third law states that for every action there is an equal, opposite, and co-linear reaction. Since anything that happens in the material world involves *interaction* between particular entities, Newton's laws of motion appeared to enable one to predict the future given sufficient knowledge of the present.

Newton was a devout if unorthodox believer; indeed, much of his life was devoted to the study of theology. He wrote *Principia* with the fervent belief that he was gaining insight into the mind of God. So it is ironic that his work gave a big boost to the secular humanism that had re-emerged in the Renaissance and gained momentum with the Enlightenment. Pierre-Simon Laplace (1749-1827), the French mathematician whose monumental contributions rivaled and extended Newton's own, set the tone in his famous quip to Napoleon: when the emperor asked him why he had not mentioned God in his masterpiece on Celestial Mechanics, Laplace replied "Sire, I had no need for that hypothesis." God (and by implication *mind*) was no longer required to explain the day to day workings of nature, only its primordial inception. From thence emerged the Enlightenment myth of Deism.

Newton's deterministic laws and calculus, along with that developed independently by his contemporary Gottfried Leibniz (1646-1716), fit beautifully with the empiricist agenda of Bacon and the dualistic metaphysic of Descartes. And since in this scheme any change of trajectory can only be attributed to perturbation by an external force, Newton's laws also fortified the Aristotelian injunction against a thing being its own cause (Juarrero, 1999). And so was born the 'scientific' notion that any material occurrence can be fully attributed to a *mechanistic* cause. The universe, in this myth, is a machine, nothing more than clockwork: everything is mechanically predetermined, so free will is an illusion. All that is needed to predict what will happen forever into the future is knowledge of the positions and momentums of all particles—knowledge equivalent to that held by an omniscient being, such as Newton's God, or Laplace's Demon.

With their combined metaphors, metaphysics, and math, Bacon, Descartes, and Newton succeeded in fully 'externalizing' the idea of causality. From this perspective it appears that the cause of something can be discovered by dissecting that thing into its component parts and figuring out how those parts interact with one another—it is simply a matter of lawful action and reaction. In this view subjective agency is merely a bothersome bias, a nuisance that gets in the way of progress toward ever increasing knowledge. Gaining knowledge requires only that the world be parsed into subjective self and objective other. And since subjective self is usually reserved for 'me and those like me', it requires only a short step of *ex post facto* rationalization to move from 'knowledge is power' to 'might makes right'.

To be sure, numerous discoveries since the time of Newton and Laplace have placed thorns in the side of determinism and its methodological handmaiden reductionism: the second law of thermodynamics, quantum mechanics, and chaos theory all suggest that the world is fundamentally un-clock-like. But within the Baconian-Cartesian framework, such indeterminacy can always be rationalized as mere epistemological uncertainty, i.e., lack of sufficient knowledge: either we have not acquired enough empirical data, or our mental model is simply not large enough.

Lest we be misconstrued, the point of this discussion is not to be anti-science. On the contrary—we need science, now more than ever. But science cannot succeed in its ostensive purpose of producing realistic knowledge without acknowledging and taking into account its cognitive limitations and metaphysical (metaphorically-based) assumptions. Otherwise it breeds hubris, which as anyone versed in Greek mythology or Mary Shelley's *Frankenstein* knows, leads to ruination. The point is that science, being a cognitive discourse, is as metaphorically bound as any other cognitive discourse.

Science is often thought of as being antagonistic to religion, because it undermines literal interpretations of religious metaphors. But as noted above, at a deeper level science and religion are really in cahoots: a deal, negotiated by the metaphysical pronouncements of Bacon and Descartes and the mathematical genius of Newton, that reinforces polarization by working to the advantage of entrenched literalists on *both* sides of the science-religion culture war. Within the Baconian-Cartesian-Newtonian framework that still holds sway, science cannot possibly 'win' the ultimate existential argument, because mechanisms by definition require an external cause. The irony is that Bacon, Descartes, and Newton all knew this to be true. The concepts of final cause and subjective mind, ceded to theology some 400 years ago, are essential for explanatory closure in the real world. If they are not brought in to our discourse on nature then any attempt to explain reality leads to infinite causal regress (not to mention nihilism) that can be truncated only by invoking the *super*natural.

And yet the belabored mythology of mechanisms lives on (Haken, Karlqvist & Svedin, 1993). Remarkably, the scientific discipline that embraces it most tenaciously is (as anyone with an intuitive feel for life itself knows) the one for which it is least appropriate: biology. As a result, science has misconceived life, and continues to do so. This despite the fact that it has been shown, with the rigor of formal logic, that 'mechanism' is wholly inadequate as a metaphor for life (Rosen, 1985, 1991, 2000). That demonstration was articulated by theoretical biologist Robert Rosen, to whom we now turn.

# Chapter 3

# ROBERT ROSEN'S INSIGHT INTO LIFE ITSELF

It is sometimes joked that biology is the science for those who lack the aptitude for math. There is some truth to this: research in contemporary biology (especially biomedical research, which receives the lion's share of funding) is largely concerned with testing hypotheses informed by empirical observation and intuition. For the past two centuries, biologists' intuitions have been entrained to the idea that causation is fully explained by mechanisms, which allows them to think that their reductionist dissections are adequate to the task of understanding life. Certainly there have been mathematicians who have developed theory that pushed the field forward, such as Ronald Fisher, who along with Sewall Wright and J. B. S. Haldane created the mathematical framework for evolutionary genetics, and Nicolas Rashevsky and Alan Turing, whose seminal work in the field of nonlinear dynamics provided the first real insight into spontaneous pattern formation ('epigenesis'), a problem that had mystified developmental biologists since the time of Aristotle. But these advances were (and still are) peripheral to the attention of most biologists, who tend to be hardcore empiricists unreceptive to theory beyond that articulated by Darwin—which (fittingly enough) was done largely without math.

Robert Rosen (1934-1998) was exceptional in many ways. Like his mentor Rashevsky, Rosen was a 'mathematical biologist', and was in fact an extraordinarily gifted mathematician. But unlike other mathematical biologists, who by and large have sought to understand biology in terms of physical dynamics, Rosen approached the problem of life in a refreshingly new way: by formally addressing the question of what it actually means to be *alive*. He was thus led to unique insights that very few scientists have yet come to appreciate—including the realization that life itself can never be fully defined by math alone, or computed. It is often noted that biologists tend toward 'physics envy', a wistful desire to be able to construct mathematical (or computational) models of biological systems with as much real-world predictive power as those developed by physicists. But Rosen's work shows us that this is wrongheaded—if anything, physicists should be looking to biology for deeper insight into reality.

Indeed, Rosen's work gets us right to the heart of the reality check that the title of this book implies is needed, a problem that we argued in the last chapter extends directly from the Reductionist way of *knowing*. The problem, in a nutshell, is that we fail to distinguish between our models and the actual world. To the extent that it exists at all, the physics envy of biologists is symptomatic of this larger problem, which is also evident in the fact that the various sciences are often pigeonholed as being 'hard' or 'soft' (with the normative implication that 'hard', meaning deterministic, quantitative, and/or predictive—as in physics—is better than 'soft', meaning probabilistic, qualitative and/or descriptive—as in psychology or even biology). In the same vein, science, being ostensibly objective, is more highly valued than the humanities, which are held to be 'merely' subjective. With respect to this mindset, Rosen took inspiration from Robert Maynard Hutchins (Mikulecky, 2011), an educational philosopher who as president and chancellor of the University of Chicago sought, through dialog, to eliminate such biases, which he saw as damaging to discourse. The critical importance of educational reform is a subject that we will take up in Chapter 7. For the moment suffice it to say that Rosen's work begins to reveal why the normative segregation of knowledge in academia, enforced by well-developed cognitive barriers engendered by Reductionism, is so damaging.

Our contention that the mythology of mechanisms—encompassing the Baconian myth of objectivity, the Cartesian myth of mind/body dualism, and the Newtonian myth of mechanistic determinism—misrepresents life follows logically from a formalism that Rosen developed and articulated in a series of technical papers and books (e.g., Rosen, 1958, 1972, 1973, 1975, 1985, 1986 a, b), culminating in *Life Itself: a Comprehensive Inquiry into the Nature, Origin, and Fabrication of Life* (1991), and its sequel (published posthumously) *Essays on Life Itself* (2000). To derive his proof Rosen used, and further developed, an advanced branch of mathematics known as category theory. His achievement resonates with that of Gödel in its demonstration that no self-consistent formal model is capable of fully representing reality. Rosen's deep insight into this problem provides a framework for weaving the various strands of thought that we have developed thus far into a coherent whole.

But first we make explicit what up to this point was only implied: our thesis that life—which might conceivably include, as we will argue, the material universe as a whole—is inherently subjective and teleological. In other words, *to live is to experience and intend*. It is important at the outset that we clarify what we mean by this, because the words 'subjective', 'experience', 'teleology', and 'intent' are loaded, having historically acquired connotations (e.g., supernatural or theological) that are definitely *not* what we mean. For that reason Rosen explicitly distinguished his work from "teleology". But at root the latter term simply refers to final cause (purpose), as defined in the last chapter: viz., the *need* that motivates action directed toward its fulfillment. So rather than throwing the baby out with the bathwater and inventing new terms, we will use these, because when stripped of their baggage they are perfectly good words that refer precisely to what we *do* mean, which should become clear in what follows.

Even so, our thesis will undoubtedly provoke strong reactions from many (perhaps most) scientists and scientifically-minded rationalists, as it undermines assumptions that Bacon encoded into the Scientific Method. It also undermines Cartesian dualism, because subjectivity connotes mind. Mind and matter are thus seen to be inseparable, manifesting different aspects of the same living reality.

Before we show how this follows from Rosen's work, we will set the stage by exploring how it is implicitly anticipated by two relatively uncontroversial scientific ideas: Darwin's theory of evolution by means of natural selection, and the laws of thermodynamics. Although many scientists may be loath to admit it, both of these ideas begin to address the question *why*, a question intentionally eschewed by reductionists entrained in the Baconian-Cartesian-Newtonian mindset.

### Teleonomic Constraint: Evolved Functionality

Darwin's theory holds that life evolves because organisms are driven to reproduce to the limits of their environment's carrying capacity, engendering competition that selectively favors those random variants that are most adept at exploiting their circumstances in order to reproduce. *Telos* is implicit in the "struggle for life" referred to in the subtitle of *On the Origin of Species* (Darwin, 1859), and is thus brought back into science. For if an organism is to survive and reproduce, its anatomy and physiology must function towards a defined end, i.e., fulfill a specific need. For example, a heart pumps blood *in order to* distribute oxygen throughout the tissues of the body, because tissues need the energy released by oxidation of food *in order to* continue living. If these needs are not met then the organism dies; so we can say the final cause of the circulatory system is the need to continue living. Furthermore, subjectivity is implied since that need is inherently selfish. Mere objects do not have selfish needs that they strive to meet; nor do dynamically deterministic mechanisms involving interacting particles animated entirely by external forces.

Stephen Jay Gould (2002) wrote extensively about how evolutionary biology is fundamentally different from other sciences. Conventional wisdom holds that science is all about testability through controlled experimentation—the Baconian ideal. But Darwin took his lead from his friend and mentor Charles Lyell, the geologist who opened science up to the investigation of history, which is closed to experimentation. History is nonetheless open to scientific (i.e., testable) hypotheses, because it is safe to assume (to a point) that things worked in the past much as they do now. Moreover, as Gould argued, empirical verification can be accomplished through the convergence of evidence from multiple independent sources. Such convergence has firmly established the empirical fact of biological evolution.

But Darwin's idea introduces a view of the world that is fundamentally non-mechanistic. Unlike dynamically interacting inanimate particles, or even complicated collections of dynamically interacting inanimate particles, organisms are causally *complex* (a somewhat ambiguous term whose meaning we discuss at length below). By placing the organism—a subjective self that strives to survive and reproduce—at the center of attention, Darwin steered science onto a new track, one that led away from the mythology of mechanisms and toward the recently emergent discourse on 'complex systems'.

And indeed, biology is all about *systemhood*—a concept that Rosen (1989, 1986a, b) developed formally with great care—as epitomized by an organism, which embodies the non-reductionist idea that the whole is greater than the sum of its parts. Although many biologists lost sight of this obvious fact with the advent of 'molecular biology' (a field, created by physicists, that greatly invigorated the mythology of mechanisms) the notion that organisms are *self*-perpetuating and *self*-regulating, the central concern of physiology, still permeates biological thought. Thus, a key biological concept is physiological homeostasis, a dynamic system's robust stability (resistance to some kinds of change) under fluctuating conditions. To maintain homeostasis, organisms undergo adaptive internal changes *in order to* persist (i.e., stay 'the same') in the face of unpredictably changing external circumstances—a reality that is nothing if not teleological.

To be sure, an organism contains within itself myriad self-organizing molecular and cellular configurations, cybernetic processes, and redundancies that allow (are required for) it to maintain homeostasis. But such physiological mechanisms come to exist because of systemic entrainment toward some end, not the other way around. As we shall see, living systems are complex: unlike the dynamic systems modeled by physics, they are *closed* to efficient cause (meaning that their *vitality* comes from within), so it is both inappropriate and misleading to think of them as Baconian objects, Cartesian mechanisms, or deterministic systems of dynamically interacting Newtonian particles motivated entirely by external forces.

Despite their propensity to maintain homeostasis, biological systems evolve with time. How can these seemingly contradictory facts of life be reconciled? Darwin provides a brilliant answer: life evolves because organisms are far from perfect. Unlike mechanisms, they are not fully determined; rather, they are (like any formal system) by nature *incomplete* (Deacon, 2011; Longo & Montévil, 2012). As much as mature organisms 'resist' change, the anatomy and physiology that works to that end is constructed by way of development, a continuously self-organizing trajectory of system-level change. As we will discuss at length in the next chapter, development produces specific variations on a general theme directed toward functionally defined yet somewhat indeterminate ends (e.g., photosynthesis, herbivory, or carnivory). To the extent that such variation is heritably encoded within the system, natural selection directs change within populations of organisms.

So organisms can be viewed a complex systems that "struggle for life" by (self)-organizing stress responses that enable their homeostatic persistence in a continuously changing world. But they evolve different forms over time because some variants are better at persisting and reproducing than others in whatever circumstances they happen to find (or create for!) themselves. In other words, organisms *strive* to reduce or mitigate stress with varying degrees of success within different environments. Evolution then is an anticipatory process of trial and error (note the normative subjectivity implicit in the latter term) driven by need, and life as we know it (in retrospect) represents those trials that were successful: natural selection is a filter through which pass only those systems that succeed in meeting the circumstantial challenges of living. So to the extent that life is worth living, it pays to be prepared.

But as Gould argued at length in *The Structure of Evolutionary Theory* (2002), natural selection is much more than a filter: it is a 'sculptor' of adaptive design. Because biological systems span multiple hierarchical levels of scale and organizational integration, each of which has heritable encodings, configurations naturally selected for a specific function at one level might produce, as a byproduct, structures at another level that (at that level) are immediately functionless, but which nevertheless provide a reservoir of adaptive potential as new needs arise. So natural selection occurs at multiple hierarchical levels of scale that are dynamically uncoupled (Salthe, 1993) and hence somewhat independent. As a result, a living system evolves at different rates at many different levels of scale and organization.

The finality implicit in Darwinian evolution has been acknowledged by scientific luminaries such as Ernst Mayr and Jacques Monod, who used to term 'teleonomic' to distinguish biological *function* from the 'teleological' concept of *purpose*. But the motivation for this semantic hairsplitting was rhetorical rather than scientific. For as noted above, Bacon had banished final cause from science, relegating it to theology—a metaphysically arbitrary (albeit historically and politically understandable) act that cemented the religious connotations with which teleological thought had become imbued in the Middle Ages. That cultural baggage was not part of the teleology originally articulated by Aristotle. In essence, 'function' and 'purpose' are synonyms for Aristotelian final cause.

Evolution thus presents a conundrum for contemporary science. On the one hand biologists (particularly those of a biomedical bent) seek to explain life by reducing it to molecular and cellular mechanisms ('determinants'). Toward that end they study a small handful of genetically-defined, experimentally tractable model organisms under highly controlled laboratory conditions designed to minimize any contextual influence, with the hope of discovering context-independent molecular mechanisms that are fundamental to life itself. On the other hand, Darwinian evolution requires, as the source of its context-dependent (i.e., adaptive) creativity, copious undirected (i.e., random) variation, an idea that is difficult if not impossible to reconcile with the idea of strict mechanistic determinism. In the Darwinian framework, causation is teleonomic rather than mechanistic: deterministic dynamics are only relevant to the extent that they enhance fitness, which in some circumstances is actually better served by indeterminism. Hence, contemporary biology skirts an uncomfortable (albeit largely unacknowledged) conceptual schism.

Monod sought to explain the problem away in *Chance and Necessity* (1972), a modern dualism that attributes teleonomic function to the 'invisible hand' of natural selection, a mindless mechanism that entrains the otherwise *random* activities of molecules interacting deterministically in accord with the Newtonian laws of motion. While most biologists are comfortable with this, it has the significant problem of maintaining an unbridgeable dichotomy between life and non-life, since natural selection only works on living systems (organisms) that strive to reproduce because their ancestors successfully strove to reproduce. The natural *origin* of life itself thus remains both unintelligible and scientifically inexplicable, except as an extremely unlikely accident of history—something not lost on savvy sophists in the 'creationist' camp.

## Teleomatic Change: Going with the Flow

The physical counterpart to the biological theory of evolution, i.e., the model developed by physicists and chemists that implicitly addresses the question 'why', is thermodynamics. In this perspective change is seen to be governed by two laws. The first says that matter and energy cannot be created or destroyed, only transformed: nothing comes from nothing. The second law says that any transformation results in an unrecoverable loss of potential: nothing comes for free. These laws make intelligible the directionality of time, and by them whatever happens can be said to happen *because* it consumes free energy (i.e., produces a net increase in global entropy); or said another way, things happen because the *potential* exists for them to happen. Here again *telos* is brought back into science, as the actualization of potential (consumption of free energy via flow directed down a potential gradient) can be construed as a 'goal'. Thus, the entropy increase mandated by the second law of thermodynamics amounts to a 'purposeful' reduction of stress (dispersion of stored potential energy), reminiscent of what organisms do in order to maintain homeostasis. In other words, in order to persist, the world apparently *needs* to reduce stress by consuming free energy. The spontaneous end-directed change that occurs by virtue of the Second Law (i.e., diffusion) is often referred to as 'teleomatic'.

Recall that the final cause of something, as conceived by Aristotle, is simply the need that it fulfills. Thus, the universal need to consume free energy ('dissipate') can be thought of as the final cause of the 'struggle for life' driving Darwinian evolution (Salthe, 1993; Schneider & Kay, 1994; Schneider & Sagan, 2005). Moreover, the randomizing effect of the Second Law produces the spontaneous mutations that provide raw material for natural selection. The problem, of course, is that while the Second Law of thermodynamics as conventionally conceived (and as encoded in the formalism of Boltzmann) handily explains why Humpty Dumpty could not be put back together again, it does not provide an obvious answer as to why Humpty Dumpty came into existence in the first place. But perhaps that merely reflects the limits of the formal context in which chemists and physicists have framed thermodynamics (Ulanowicz, 2009b).

In any case, from a thermodynamic perspective (even as conventionally framed) anything that can be construed as a 'problem' that needs solving can also be reasonably construed as a constraint that impedes flow along some path of potential. So to the extent that science and engineering work to solve problems, they work to decrease *resistance* by removing such impediments. Reductionism is well-designed to do just that, because it narrowly focuses the acquisition of knowledge on mechanisms, which are essentially metastable kinetic constraints on flow. As discussed in the last chapter such knowledge grants power, which (according to Bacon) is its purpose. But that narrow focus always produces unintended consequences, because of the *lack of attention* to potential that is irrecoverably lost every time a mechanistic impediment is overcome by human engineering. Owing to the Second Law, reductionist science that is not constrained by humanity is inherently (albeit unwittingly) nihilistic.

From the foregoing it should be apparent that 'science' is anything but monolithic in its worldview. The prevailing mythology of mechanisms co-exists with well-developed scientific ideas that implicitly undermine its founding propositions. And yet science as currently conceived is widely considered to the only means we have of acquiring realistic knowledge. From a philosophical perspective this is problematic. But since philosophy fell out of fashion with the ascendancy of science, most scientists fail to see the problem, and hence have no problem keeping a straight face while advocating an 'objectively realistic' worldview constructed from knowledge obtained through a multifaceted lens of inconsistent models.

The widespread aversion to philosophy among scientists is understandable however, given that scientists function to perpetuate a larger *system* that in turn entrains their activities. Bacon's declaration that the purpose of knowledge is to increase human control of nature placed science in the service of technology. Thus enabled, technology increases economic power, which in turn forms the basis of political power. Politics then selects what science is to be economically favored, closing the loop. This system has no 'use' for the penetrating questions of philosophers, which only impede economic development by casting doubt on the mythological rationalizations that are culturally propagated to sustain the system's technologically-enabled manipulation of nature. As a result of this selection pressure, the system favors the evolution of cognitive defenses—including both anti-intellectual populist and narrowly-focused academic mindsets—that serve to thwart any philosophical threats to its persistence. It is no wonder that philosophy is widely considered to be a useless and highly esoteric occupation of a small number of eccentric academics.

So, to be fair, science in its current stage of development does not afford the capacity to comprehend final cause, because scientists remain committed to the Baconian-Cartesian stricture against bringing subjectivity (mind) into explanations of nature. As a result the scientific concept of causation has collapsed to strictly connote a temporally 'linear' sequence of direct (i.e., mechanistic) cause and effect. The acknowledged irreversibility of time (attributable to the Second law of thermodynamics) would then appear to rule out any possibility that an effect can be part of its own cause, thus reifying Aristotle's injunction against self-cause (Juarrero, 1999). Such circularity makes no sense within the mental frame entrained by the mythology of mechanisms.

It does however make complete sense to a *mind* that *anticipates* the future outcomes of actions that it directs. Human beings have no problem identifying a final cause for their own activities. We build a house because we mentally anticipate a need for shelter. The problem then is that the subjective agency of mind—that which affords the capacity for anticipation and intentional action—is thought to be uniquely human, an 'emergent' (or if you prefer, 'God-given') property of our extraordinarily large brains. Although you would be hard-pressed to find anyone who can provide a convincing argument for why that would have to be the case (much less hard evidence that actually supports the proposition), this metaphysical legacy of Rene Descartes, his concession to the seventeenth century Church, lives on, largely unquestioned by science.

The notion that human beings are the sole beneficiaries of mind is not without irony. For as discussed in the first chapter, our technological wonders have come at great expense, resulting in an existential threat to civilization that was all too predictable given reductionism's nihilistic bent. Apparently the vaunted mental acumen of *Homo sapiens* leaves something to be desired. Maybe we do not differ from our animal cousins as much as we like to think. Could it be that the widespread perception that we are the only species that *acts with intent* is more delusional than real?

The fact of the matter is that life itself is anticipatory. Organisms could not survive and reproduce without anticipating future needs and working to ensure that those needs are fulfilled *before* they become life-threatening. In recognition of this fact Rosen coined the term 'anticipatory systems' (Rosen, 1985). His definition of what this means makes explicit what most people instinctively know to be true: that life entails mind (Thompson, 2007). This follows from what Rosen discovered by asking what must happen for us (or any living system) to *know* something about the world. In other words, what are the minimum formal requirements for the acquisition of knowledge? The answer is 'the Modeling Relation'.

## The Modeling Relation

In order to anticipate its future needs, a system must become informed about the world as it *relates* to those needs; that is, it must acquire knowledge about *relevant* aspects of the world. Contrary to what is perhaps common sense, knowledge acquisition does not occur by passive absorption of 'information'. Rather, it occurs by way of experience, which entails *attention*, a form of intentionally focused work. The key to understanding why this must be so is found above in the word 'relevant': the world has myriad (indeed, an infinite number of potentially informative) aspects, only a small subset of which is relevant to the needs of a system. Those aspects that are *most* relevant are those that predictably produce an outcome that satisfies the system's needs. That which is predictable in nature is that which is caused. So to acquire knowledge about the world as it relates to its needs, a system must become to some extent informed about causation.

The only possible way to do that is to construct a formal model. Toward that end a system must semantically *encode* relevant aspects of the world, and then determine, through an internally embodied system of syntactically formal logic, what those encodings *entail*. Such entailment need not represent any reality beyond the system itself, and thus may not have anything else to do with the real world. However, when *decoded* back into the world by way of the system's actions it may well fit the system's needs to its circumstances, in which case a model is born. Rosen (1985, 1991) referred to this anticipatory loop as the Modeling Relation.

The significance of the last step in establishing the relation, the act of decoding, cannot be overstated. It is only through decoding that the results produced by an internal system of entailment become externalized. This is fundamentally a *creative* act—in other words, it is *art*. If the decoded creation fits (is 'congruent' with) an 'external' aspect of reality, the Modeling Relation is said to *commute*—a situation wherein one system (e.g., an organism) comes to embody a model of a second system (e.g., an ecological niche), which is in turn a *realization* of the model. So when the Modeling Relation between two systems commutes, reality is not just anticipated within the system doing the encoding and decoding, it is actively created within the world at large. In this way ecological niches are intentionally constructed (Laland *et al.*, 1999), as for example occurs when bees construct their hives and beavers build their dams. Through the Modeling Relation prophecies can (and sometimes do) become self-fulfilling, and dreams can (and sometimes do) come true.

So the acquisition of realistic knowledge about the world is an anticipatory process that requires attention, at some level, to signs that are specifically relevant to the needs of a system, which are deciphered by the creation of models. It should be obvious that the Modeling Relation is the basis of the Scientific Method. But it goes much deeper than that. The Modeling Relation is fundamental to life itself, and makes sense of much of the foregoing discussion.

For example, it provides insight into both metaphorical and mathematical language. Metaphor, after all, is a means of semantically encoding signs—and quite a creative means at that, given that it uses one thing to describe another. Numbers are also semantic encodings. Linguistic and mathematical syntax in turn afford formal systems of entailment. And decoding occurs when semantics and syntax are put to active use in order to *create* something—a novel for example, or a house (or a beaver dam or beehive). Metaphysics and science are, in essence, cognitively creative art forms that use metaphorical and mathematical encodings, together with their syntactically constructed entailments, to develop models: they are conscious human manifestations of the Modeling Relation.

With that in mind it is interesting to revisit some of the territory covered earlier in this essay. Consider Platonic Ideals, which Plato held to be a reality separate from the material world of the senses. And in a sense they are: ideal forms (e.g., a Line) semantically encode—that is, they *define* in words—some distinct aspect of reality. The same can be said of Aristotelian principles gleaned inductively from the particular instantiations of nature. Those semantic encodings become *models* to the extent that they syntactically entail something within the system that, when decoded, is realized within in the world at large (e.g., a line drawn in the sand). Metaphysical models are thus created by the anticipatory system of human consciousness.

Now consider the circularity alluded to above, which many a reductionist finds irksome. From the Modeling Relation we see that any internal model involves a closed loop between itself and its external realization. In other words, a model develops from a 'circular' relationship between (at least) two systems. Such relationships are ubiquitous in nature, and are the basis of what any modern engineer knows to be the key to control: feedback. Positive feedback is the basis for signal amplification, or more generally, growth, which can in turn be controlled by negative feedback, allowing for regulated development. It should be apparent from the Modeling Relation that a realized model affords positive feedback. But it also affords the potential for negative feedback, as its realization depends on circumstances—that is, it is context dependent. Negative feedback occurs when the actions of an embodied model produce circumstances that negate the model's realization.

Thus, the Modeling Relation engenders, through positive and negative feedback, the construction and refinement of models, and concomitantly, the creation and development of reality—and indeed, of life itself.

But this is where the human mind reveals its limits. Well-tuned for modeling systems at its own 'focal' level, it finds itself at a loss when modeling systems that encompass hierarchical levels that are further removed, wherein the human component is less 'present' (Salthe, 1993). The world-at-large spans levels of scale and specification that lie far beyond our mind's ability to fathom in any detail. Modeling those levels requires arcane math, resource-intensive (and hence ecologically disruptive) technology for obtaining measurements, and overextended metaphors; so rather than trying to 'see it all', most of us either resort to faith in the authority of 'experts', or simply shrug our shoulders.

## Simple versus Complex Systems

With a little reflection on the matter you should be able to see how the Modeling Relation manifests in any living system. Take for example some of the 'simplest' of single-celled organisms, bacteria. A bacterium *senses* its environment: it will move toward food and away from toxins. It does this because, by way of a well-characterized process of 'signal transduction', it can encode what it encounters into molecular signs (semantics) that are subject to ('interpreted' by) a biochemical system of entailment (syntax). The entailment is then decoded as action—movement directed by the activity of a flagellum. If the direction of the movement fits with the needs of the bacterium—toward food and away from toxins—then the bacterium has modeled its reality. It simultaneously creates a new, somewhat more bacterio-centric reality, to the extent that its success enables its self-reproduction, and hence the propagation of its internal model (which is *recorded* in its DNA).

But is this not just a deterministically reactive mechanism of 'stimulus-response' attributable entirely to the dynamic 'laws of physics' that govern the interaction between particulate entities (molecules) within the bacterium? No it is not, by any stretch. Any biologist should know this. While the biochemical syntax is indeed mechanistic, no law of physics requires that the bacterium move toward food and away from toxins. With respect to physics and chemistry such movement is completely arbitrary. A bacterium 'knows' to move toward food and away from toxins because it has successfully modeled reality, and hence been selected to reproduce. The relevant explanation here is not Newtonian mechanics but Darwinian evolution, which, as discussed above, implies teleonomic (final) cause. Mechanistic (efficient) cause does not account for the fact that the vast majority of bacteria alive today successfully model those aspects of reality relevant to their survival and reproduction, because the untold numbers of dead bacteria that did *not* successfully model their reality (and hence did not reproduce) were no less mechanistic.

Consider another example: a developing embryo. An embryo epitomizes the creativity of the Modeling Relation. It begins its individual existence as a single, relatively amorphous cell—a zygote—and ends with a fully formed multicellular organism specialized for certain kinds of work—for example a human being, or an oak tree. The teleology of this developmental sequence is obvious: the purpose (or if you prefer, function) of the embryo is to produce a specific kind of organism. This purpose is fulfilled by way of the Modeling Relation. Here again the encoding is carried out by molecular signs (signals presented within the embryo itself and by its environment) that are interpreted by a biochemical system of entailment. The latter is inherited, its syntax encoded within the regulatory sequence of the DNA molecule, which affords an abstract formalism that biologists refer to as a 'gene regulatory network' (Davidson, 2006). What this network logically entails is algorithmically decoded, in part, via the 'Central Dogma' of molecular biology: through 'transcription' and 'translation' of gene sequences into protein structure, which then feeds back (via specific interactions encoded in DNA sequence) to regulate transcription and translation. The result of this progressive decoding is an unfolding 'program' of development, wherein each cell within the organism participates, by virtue of its dynamically changing protein repertoire, in the formation of a functional tissue or organ—a heart, an integument, or a brain—while becoming progressively specialized for a particular function—a blood cell for carrying oxygen (using the protein hemoglobin), a skin cell for protection (using the protein keratin), a nerve cell for communicating long distances (using various proteins that regulate electric currents), etc. To the extent that these functions fit the organism's needs to its circumstances, the Modeling Relation commutes; that is, the organism comes to embody a model, and its environment a realization thereof. The better the model, the more fit the organism.

But again, is this not a mechanistic phenomenon? The answer is again yes but (more emphatically) no. Mechanisms certainly come to play in the syntax that governs the developmental process—no physical laws are broken. But the physical laws that govern the dynamic interactions of particles do not and can not adequately account for the development of an organism from an egg. Nothing in the sciences of physics and chemistry is predictive of a human being or an oak tree. Additional scientific models are needed. Darwinian evolution is one such model, which provides a scientific framework for rationalizing mechanism (efficient cause) in terms of a function (final cause) and vice versa. Indeed, since Darwinian Theory is commonly assumed by biologists to account for function, teleonomy has become implicit in their usage of the term 'mechanism'. So when a biologist seeks to discover the 'molecular mechanism' underlying a given phenomenon, the rationale for doing so is to elucidate a function—how the mechanism fulfills the specific needs of the organism. Mechanistic accounts of function are in fact a prerequisite for obtaining research funding and publishing in the top biological journals. In the context of pure physics however it is meaningless to speak of a mechanism's function, just as for biologists it is meaningless to speak of the function (or purpose) of natural selection (the mechanism for teleonomic change). So 'mechanism' holds somewhat different meanings in different contexts. Yet it is the same word, with the same deep connotations that inform our collective cognitive framework.

Despite its success in expanding that framework, Darwinian Theory also falls short as explanation. Nothing in the theory of evolution is predictive of a human being or an oak tree. Moreover, the deterministic connotations of the term 'mechanism' become misleading when the discord between Newtonian determinism and Darwinian evolution is left unacknowledged.

This brings us to a key insight afforded by the Modeling Relation: that there can not possibly be a single largest model of the real world. This follows from the fact that a model is nothing more and nothing less than the product of a commutative *relationship* between (at least) two systems. It requires an encoding, which depends on the information capacity of the system making the model, and just as importantly, it requires decoding, which depends on the creativity of the same system. Since the information capacity and creativity of real systems (that is systems that are materially embodied and hence governed by the laws of thermodynamics) are necessarily limited, they require that choices be made. Attention to one thing must be paid *at the expense* of another. A model is thus, of necessity, an *idealization* that can at best make sense of some aspect of reality, involving subjective commitments that invariably occlude other aspects. Furthermore, there is no law of nature that requires self-consistent models to be consistent with, and hence derivable from, one another. This too follows from the fact that the Modeling

Relation entails choices, commitments to a specific 'point-of-view' that necessarily remove other (incongruous or irrelevant) aspects of reality from view.

In short, reality is *complex*, whereas models are simple (Mikulecky, 2007a, b). What this means is that reality cannot possibly be 'captured' by any single formalism. The best we can do is find different ways of interacting with the world that allow us to develop distinctly different models that are not derivable from one another. Since any 'mechanism' is an encoded concept whose entailments are realized in the world, it is a model—an idealization—of a specific aspect of reality. The same is true for Darwin's theory, and the laws of thermodynamics. Because they are models, none of these scientific ideas can give a complete explanation of the real world. Even combined they lack that capability. And while together they do better than any one model alone, as noted above combining them creates the philosophical problems of consistency. Newtonian mechanisms are reversible, whereas thermodynamics is not. Newtonian mechanisms are predictable, whereas evolution is not. In reality, countless self-consistent models that are to some extent contradictory can and do co-exist.

This explains why the mythology of mechanisms is just that: a *myth*. Laplace's demon is impossible, not simply because omniscience is impossible, but because reality involves far more than Newtonian mechanisms, and hence is not altogether predictable from the positions, momentums, and interactions of its particulars. Mechanisms, being models, are simple. Reality is complex.

Complexity is often described using the cliché 'the whole is greater than the sum of its parts'. That which remains after the parts are subtracted from the whole is referred to as 'emergent'. Thus, the whole is greater than the sum because an irreducible reality *emerges* from the specific way the parts are organized. Biology cannot be reduced to chemistry because the syntax that mediates biological entailment—its *organization*—is not completely determined by (i.e., is arbitrary with respect to) the syntax of chemistry. Attempts to reduce biology to chemistry and physics are doomed to failure, because biology manifests causative information (constraints) that chemistry and physics do not. In discussing this reality forty years ago, the Nobel-prize winning physicist Philip W. Anderson noted that the laws of physics are symmetrical, whereas life breaks symmetry in ways that are not in any way predictable from those laws (Anderson, 1972).

But complexity also implies that 'there is more than one way to skin a cat'. The idea captured by this metaphor is that functional organization does not require a specific material configuration: the same needs can be satisfied in different ways. Thus, both the past and future are largely indeterminate with respect to the present, and the parts are indeterminate with respect to the whole. The only way to learn anything about history is through the interpretation of temporally stable artifacts such as written texts, buildings, fossils, DNA sequence, and electromagnetic radiation. If the world were simply a mechanism this would not be the case, as mechanisms are reversible: to a blind, amnesiac Laplace's demon that somehow managed to nevertheless have complete knowledge of the moment, both past and future would be computationally transparent. But thermodynamics tells us that the world is both irreversible and indeterminate, rendering past and future *incomputable*.

Ironically, it was Alan Turing himself, the father of computer science who is often associated with the proposition that any real (i.e., physical) process is computable, who showed that the opposite is more likely true, by formally proving the existence of real (i.e., definable) numbers that are mechanically incomputable (Turing, 1936)—a result (if not its interpretation) praised by Gödel (Shagrir, 2006). Unfortunately, of Turing's many brilliant mathematical insights, this is the one that appears to be largely lost on the modern world of science.

Because modeling requires that *subjective choices be made about what to attend to*, mechanistic models are not derivable from thermodynamic models and vice versa (Mikulecky, 1993; Schneider & Sagan, 2005). Each of these models contains elements that render their union untenable, as the combined model would not be self-consistent. The closest that we have come to such a union is 'statistical mechanics', which simply avoids the inconsistencies by sweeping indeterminacy under the epistemological rug of 'uncertainty'. Both mechanisms and thermodynamics are nevertheless realized in the world—otherwise they would not be models, much less part of the scientific discourse. But neither gives a full and accurate account of reality—no model can.

So once again: causation in the real world is complex, whereas the models that we used to explain it are simple. We need all four of Aristotle's causal categories because they model different aspects of the world that are not formally derivable from one another. Bacon's choice to reject final cause led to the development of an over-simplified worldview, because his rule that science is only concerned with 'how' and has no business asking the question 'why' (one of the first questions we all begin asking as children!) produces an impoverished entailment structure that is incapable of modeling fundamental realities. This made it easier to focus on technology development while keeping the faith that science could ultimately develop a mechanistic model congruent with all of reality. As a result of that idealistic 'tunnel vision', the mechanistic model became increasingly realized (through its commutative decoding into technology) in an anthropocentrically mechanized world with little regard for the non-mechanistic mess created by that realization. Beyond the mess itself, the belief that life itself is bereft of intrinsic meaning, purpose, and value led predictably to the polarization of belief systems that we see today: existential nihilism on the one hand and supernaturalism (or superstition) on the other.

From the foregoing it should be apparent that 'complex', as defined by Rosen and as used here, is not synonymous with 'complicated'. If the world were merely complicated it could be described in a single largest model, which could be decomposed into additive (complementary and non-contradictory) sub-models. But this is not the case. The complexity of the actual world, revealed in the fact that models derive from the Modeling Relation, means that no model, no matter how large, can possibly represent reality. Moreover, any model is an *idealization* that contradicts other idealizations that model some other aspect of the world. So a single model may for a time succeed, through its embodied realization, in entraining a reality that is more congruent with itself, but this *development* will eventually fail if it does not account for aspects of reality that the model occludes—a phenomenology that we discuss at length in what follows.

Complexity explains why ideals are never fully realized—why, for example, ideas such as capitalism and communism 'look good on paper' but always fall far short in their realization. This follows from the fact that ideals are formed by models, and hence bound by the formal logic of the Modeling Relation and the subjectively arbitrary choices that its instantiation requires.

But from this perspective 'realism' is no less naïve than 'idealism' (and vice versa). Coming to grips with complexity (i.e., reality) requires that we accept that there are innumerable ways of viewing, describing, and interacting with the world, each way demanding the creative development of a different model that becomes, to the extent that it is realized, part of the world. Hence the world affords, and is compatible with, an infinite number of models, no single one (or subset) of which can possibly provide a comprehensive description of the world itself. It thus calls for humility.

Following Rosen, we have argued that modeling is the only means that we have at our disposal for acquiring knowledge; and furthermore, that it is impossible to fully represent reality through the lens of any one model. The problem with Reductionism, which acquires knowledge through the lens of mechanistic models, lies not with those models—clearly they are powerful and to a large extent realized, and we have no alternative except to use the best models that we can develop. Rather, the problem is that by virtue of its power, Reductionism has encouraged the belief that it can represent *all* of reality, which is unrealistic because no model possibly can.

## On the Generality of Life

What is life? This is the question that initially motivated Rosen, and countless others before him—including theoretical physicist Erwin Schrödinger, whose famous book posing that question (Schrödinger, 1944) was seen by Rosen to challenge the very foundations of physics (Rosen, 2000). In elucidating the process that is required to answer the question, Rosen (Mikulecky, 2000) provides us with a partial answer: life manifests the Modeling Relation. Every organism *embodies* a subjective model of some aspect of reality. Darwin described life as 'endless forms most beautiful'. He could as well have been describing the subjective nature of reality, and the number of self-realizing models that it engenders. And thus, in considering the question 'what is life', Rosen came to a brilliant and revolutionary insight that very few scientists are aware of, much less willing to entertain: that life itself (as opposed to our scientific model thereof, i.e., biology) is *generic* to reality (Rosen, 1991).

This stands conventional scientific wisdom completely on its head. To see how so we can employ a cognitive model that Stanley Salthe (1993, 2012) refers to as the 'specification hierarchy'. By definition, a hierarchy consists of nested sets of inclusiveness. The specification hierarchy is based on the nested relationship {generic{specific}}, where the brackets indicate that any specific thing can be cognitively assigned to a larger, more generic class of similar things—for example human beings are classified as mammals, which are classified as animals—as expressed in the specification hierarchy {animal{mammal{human being}}}. Thus, science has constructed the specification hierarchy {physics{chemistry{biology}}}, which holds that the biological world is a specific manifestation of a more generic chemical world, which is in turn a manifestation of an even more generic physical world. The problem however is that {physics{chemistry{biology}}} is a cognitive model developed by human beings—an *ideal* that does not (and can not) fully represent the actual world, and may well misrepresent life itself. The reason to suspect the latter is that physics is a mechanistic model of inanimate objects that does not (and cannot) account for the teleological character of living subjects—and the same is true for chemistry. Physics and chemistry fail to predict life, which is why the origin of life remains a major unsolved problem in science, and may well be intractable within the framework of contemporary science.

Here we can indulge in an interesting speculation. As discussed above, life on earth can be viewed as the planetary system's 'solution' to the problem of stress reduction in the face of thermodynamic disequilibrium. The complex of geosphere, biosphere, and atmosphere is a homeostatic system. Nothing that we distinguish as separate objects within that system exists apart—relationally, it is all one *being*, a living subject with its own unique character. Now the speculation: if we ran the evolutionary 'experiment' many times we might find that sometimes the system failed to stabilize and this unique character would not emerge. But in those cases where it did, its *particulars* would never manifest in exactly the same way twice. If this is correct (and we think it is), there can never be the one to one mapping between the system and its parts implied by mechanistic conceptions. Particulars are causally relevant (i.e., meaningful) only in the context of the global system. This is what is entirely lost on reductionist science, and hence on contemporary secular civilization, which looks to such science for solutions to real-world problems.

The problem then is that modern civilization fails to distinguish its models from reality. We might thus question whether life itself is *really just* a more specified form of chemistry. To begin to see the value of this question consider that 'biology' (a cognitive model of life as we *know* it) is not synonymous with 'life'—and that functional organization does not depend on a specific material configuration. Biology is a formal model of those aspects of life that are cognitively accessible to science as currently framed, viz., as everything in the universe being a specific outgrowth of mathematically-tractable physics. The problem with this, according to Rosen (2000), is that:

> *The conventions on which contemporary physics rest amount to asserting that the world of material nature, the world of causal entailment, is a predicative world. It is a world of context-independent elements, with a few finitary operations, in which impredicativities cannot arise. This is what makes it objective. But this objectivity is bought very dearly: its cost is a profound nongenericity of that world, an impoverishment of what can be entailed in it. Most profoundly, it is a world in which life does not exist. But life does exist. The world of material nature is thus not in fact a predicative world. That is the Foundational Crisis faced by contemporary physics—that the world to which it aspires is a complex one, not a simple one. (p. 44)*

Hence, it appears that the scientific frame—{physics{chemistry{biology}}}—is too limited, and hence too limiting, to model *life itself*. Consider that *science itself* is but a specific expression of life, which gives us {life{science{...}}}. Framing the conventional scientific model in this context then gives us {life{science{physics{chemistry{biology}}}}}. From this perspective it is quite conceivable that life is actually generic to material existence—that our failure to recognize this may simply reflect the epistemological subjectivity inherent in the Modeling Relation, and the particular choices made during the historical realization of the mechanistic scientific model that produced {physics{chemistry{biology}}}. Might the 'origin of life' actually coincide with the 'origin of the universe'? If life itself can be so conceived, where might that lead us? Might such a notion (i.e., 'hylozoism'), when decoded, somehow commute with the world at large and thus become realized as a model? And might that offer a way out of some of the dire predicaments faced by contemporary civilization?

Rosen's elucidation of the Modeling Relation suggests that the question 'what is life' is in a sense ill-posed (Mikulecky, 2000), at least from the perspective of contemporary science. A more scientifically tractable question is: how does life differ from a mechanism? In addressing that question Rosen found that life differs fundamentally from the standard mechanistic models of science—that unlike any mechanism, life is self-referential and self-entailing. The model used by Rosen to show this involves two processes: metabolism and repair. Metabolism is the process—the set of efficient causes—whereby one material thing is transformed into another. The material entities (e.g., enzymes) that mediate the transformation are, like everything, transient, and so must be continuously replaced (or 'regenerated') by a system of repair. The material entities that embody the latter system are in turn a product of metabolism. Metabolism and repair thus constitute a closed causal loop, a self-entailing system.

One might surmise that any material configuration that embodies this relationship could be considered to be alive; and conversely, that any living system embodies some version of the relationship. To gain further insight into life itself we might thus ask: what engenders metabolism and repair? And more generally, how are models created? To answer those questions we need to consider how cycles of interdependency come to exist and what happens when they do (Kercel, 2007; Louie, 2007, 2009; Mikulecky, 2010, 2011; Ulanowicz, 1997, 2009). It is to these questions that we now turn.

# Chapter 4

# THE LOGIC OF DEVELOPMENT

The word 'development' is used in many different contexts: we speak of developing organisms, personality, habits, addictions, disease, technology, careers, land, economies, weather, stories, and photographs. That the same word can be meaningfully applied to such a vast array of different phenomena bespeaks of a commonly sensed reality that pertains to them all. Whenever we speak of *development*, we are referring to a symmetry breaking process that brings some definite thing into existence—the creation (or 'emergence') of something from (apparently) nothing.

Definite existence—the state of being one thing rather than another, something rather than nothing—implies asymmetry, because *to be* (something) implies *not to be* (something else, or nothing at all). One implies zero; yin implies yang; life implies death. So to the extent that 'development' refers to symmetry-breaking change, it refers to the process of *becoming*: definite, and hence definable; knowable, and thence known.

Before we explore what this entails, there are two important points to bear in mind. The first is this: the asymmetry of existence ('to be or not to be') means that the development of anything always comes at the expense of something else. To be(come) any specific thing is to not be(come) another. In other words, development invariably incurs a cost, a loss of potential, capacity, or opportunity. Because of this and the fact that all resources are limited (by the laws of thermodynamics), the phrase 'sustainable development' is an oxymoron: continued development of any specific thing ultimately leads to the demise of that thing. In this sense development is a trap.

This brings us to the second point, which concerns the *finality* implicit in the word 'development': to say that something is developing implies that we have some sense or knowledge beforehand of what that thing is or will ultimately be. The reason of course is that anything that can be said to develop belongs to a bigger class of similar things that are *known* to have developed—we know what's coming because we've seen it before, in one guise or another. *Development* thus evokes a sense of inevitability, or at least of predictability: a trajectory of change through which a specific fate or destiny is realized. Through development the potential for something that does not (yet) materially exist becomes progressively actualized into the existence of that thing. The logic of development is thus the logic of *telos*—i.e., *teleology*.

In light of the discussion in the last chapter it should be coming apparent that a parallel exists between development and the Second Law of thermodynamics. Both are finalistic concepts that imply loss of capacity or potential. But whenever we think of developing something (technology or land for example) we tend to focus on the accrued benefits while ignoring ('externalizing') the cost. We submit that this tendency is reinforced by reductionism.

On this point it is worth reprising what has come before in order bring into sharper focus the widespread, but widely unrecognized, misperception wrought by reductionist science. As we have argued, science (and more generally, life itself) is all about modeling, and thus harnessing, the causal bases for nature's predictability. The natural philosophers of seventeenth and eighteenth century European Enlightenment made such great strides in developing mechanistic models that they, and as a consequence we, were led to believe that predictable change signifies—and *only* signifies—the efficient cause of dynamically deterministic mechanisms. This belief is epitomized by Laplace's famous quip to Napoleon. As a result, the predictable occurrence of anything is now commonly interpreted to be both explainable, and adequately explained, by the mechanisms to which it can be reduced.

But as we have seen there are serious problems with adopting this view, not the least of which is the fact that, being a model, it cannot possibly be the whole story.

But how might we explain predictability if not by way of mechanisms?

The answer is: *by way of development*.

This statement will no doubt provoke a knee-jerk response: isn't that putting the cart before the horse? After all, the scientific paradigm of organismal development that has prevailed for the past century and a half, known as *Entwicklungsmeckanik* (developmental mechanics), derives its name from the German embryologists who sought to account for the predictability of ontogeny by identifying its underlying mechanisms. From this perspective development merely describes the *effect* of a set of mechanisms, and is not in any sense causative.

But this is based on the Aristotelian assumption that an effect cannot be its own cause (Juarrero, 1999). As Rosen has shown, this assumption is problematic, and in fact does not apply to life itself, which manifests circular (self-entailing or reflexive) causality. Metabolism persists be*cause* it maintains a system of repair that persists be*cause* of metabolism. The chicken constructs an egg, and vice versa.

Moreover, as we have seen, causality is complex. A common over-simplifying tendency encouraged by reductionism is to seek THE cause of something. In the scientific field of developmental biology, for example, the cause of a specific developmental process—e.g., formation of an eye—is assumed to be elucidated when a mechanism is found to be both 'necessary and sufficient' for completion of that process. Thus, a specific gene might be found to be *necessary* for eye development, as shown by a failure of eye development when that gene is removed or mutated. The same gene would then be said to be *sufficient* for eye development if abnormal ('ectopic') activation of that gene leads to development of an extra eye within a tissue where that does not normally occur (e.g., a limb). Such genes are often referred to as 'master control genes'.

The problem with referring to a mechanism as 'necessary and sufficient' is that it implies something that is not true: that a developmental process is adequately explained by the mechanism. Although most developmental biologists, when pressed, would probably admit that there is more to it than that, many (perhaps most) fail to see that a 'necessary and sufficient' mechanism identified within a highly controlled experimental context is not the same thing as a 'necessary and sufficient' cause. Hence, problems of experimental reproducibility that arise when the same process is studied in different contexts (e.g., as when the effect of a given genetic perturbation differs in different strains of mice, or in different laboratories, or even in the same laboratory with cryptically different conditions) often come as a surprise and are a source of great consternation, and more often than not are swept under the rug of 'experimental error', rather than studied to learn what such variation might be telling us about causality.

Within this mindset, models of causation become framed as binary, 'either-or' propositions. But as we have seen this is unrealistic, since real-world complexity elicits multiple models that are each realized in some way, but which nonetheless evoke incompatible *ideals*. The reductionist mindset occludes context. But in complex systems context is causative because it constrains, and thus predisposes, what can possibly occur. And as we will see, development, being a definitive system-level process, establishes context.

So perhaps we can turn *Entwicklungsmechanik* on its head and ask whether, and to what extent, mechanisms are caused by development. That is to say: are mechanisms, whenever and wherever they are realized in the world, in some sense the predictable culmination of development?

This is obviously true for those mechanisms that constitute the anatomy of an organism (an eye for example, or a heart), each of which develops into existence from unformed precursors. But from the reductionist perspective of most developmental biologists, this is because—and *only* because—underlying molecular-genetic and cellular mechanisms are 'necessary and sufficient' to deterministically direct that development. And so the development of one set of mechanisms (anatomical) is attributed to—presumed to be caused by—the activities of another set of much smaller mechanisms (molecular). How the precursors of the latter mechanisms came to be in the primordial past is a question asked only by the relatively small cluster of biologists, chemists, and physicists who work on the origin of life problem, which is the major unsolved problem of science. But even if and when this problem is solved mechanistically, that is still merely attributing one set of mechanisms to another. As we have already discussed, this will always be the case in reductionist science, which leads any attempt to explain origins into infinite causal regress. But there are other problems with attributing development to mechanisms.

Recall that Darwinian evolution requires copious undirected variation. So if organismal development is entirely attributable to molecular mechanisms, then organismal variety would map simply to variation in those mechanisms. Conversely, similarities between organisms would require similarities in mechanisms. But from everything we have learned over the past three decades, the opposite is true—variation in organismal phenotype does not necessitate variation in molecular mechanisms (which are generally conserved, and often functionally interchangeable, between highly divergent life forms), and *vice versa*. Mechanisms at each level of scale (molecular, cellular, and anatomical) are, to some and perhaps a very large extent, uncoupled. In other words, organismal anatomy and physiology are somewhat indeterminate with respect to molecular biology. If it were not so we could not use nematode worms as a 'model system' for biomedical research directed at understanding the molecular biology of human health and disease—the mechanisms would be too divergent. Moreover, the fact that two different species of nematode with similar anatomy actually develop by way of different 'necessary and sufficient' molecular mechanisms—a phenomenon that evolutionary developmental biologists have referred to as 'developmental system drift' (Wang & Sommer, 2011)—is inexplicable in terms of strict mechanistic causation.

The reason that development is mechanistically indeterminate is that as far as natural selection is concerned, what matters is not *how* something works, but *that* it works. Mechanisms are naught but a means to an end (reproduction), and there is always "more than one way to skin a cat".

Nevertheless, molecular mechanisms clearly play a critical role in organismal development. To get a better perspective on the problem it might help to distance ourselves, at least to begin with, from what most biologists mean by the word 'development': a synonym for organismal ontogeny or embryogenesis, which is now widely understood to be the pre-programmed outcome of an algorithmic process encoded by genetic (or 'genomic') information. For to be sure, molecular mechanisms are a key aspect of this process, which does not work properly if those mechanisms are perturbed, and in fact can be directed toward a predictably different end by specific molecular-genetic perturbations. Our mechanistic model of development is thus very neatly realized during the development of an embryo, which makes it difficult to use this cognitive frame as a point of departure for imagining how development might itself be in any way causative.

From an evolutionary perspective however it is plausible that the underlying mechanisms that direct ontogeny are themselves the products of some sort of development that occurred much earlier in the history of life on earth. It makes sense to say (and indeed it is often said) that as life has evolved it has also developed, and vice versa; that is, to speak of 'evolutionary development'. This is not redundant because development is *not* synonymous with evolution: development refers to progressive change, constitutive of a given system, whereas evolution refers to any irreversible change (Salthe, 1993). Development implies finality; evolution does not—at least not until selection is invoked as its driver, in which case (as discussed in the last chapter) we are forced to admit that the concept of adaptive biological *function* is inherently teleological.

In differentiating development from evolution it is useful to distinguish between two even more fundamental types of change: that which occurs spontaneously (as when a car rolls downhill, or an ice cube melts in a glass of warm water), and that which requires work (as when a car is driven uphill, or a freezer makes ice). Work occurs whenever different trajectories of spontaneous change, i.e., 'flows', are asymmetrically opposed (Deacon, 2011). Development entails (and enables) work, and evolution, constrained thereby, 'goes with the flow'. Development works to create higher level organizational constraints that persist to some extent even after the lower level material constituents of the developing system disappear; as a result, successively higher levels of emergence evolve irreversibly, in ratchet-like fashion (Juarrero, 1999; Deacon, 2011), through a concatenated series of developmental cycles or trajectories (Coffman, 2006).

Natural selection then, to the extent that it directs progressive change, can thus be viewed as contributing to evolutionary development. However, as formulated by Darwin natural selection requires a generative system of inheritance, which we now know to be afforded by the DNA-based genetic system that is common to all contemporary forms of life. And with that system the apparent goal-directedness of ontogeny can easily be rationalized as a mechanistic outcome of algorithmic programming cobbled together by natural selection.

So the questions that must be asked are whether the finality implicit in 'development'—its apparent directedness towards some need-based goal—is real or illusory; and whether such directedness pertains in any way to non-genetic phenomena, which presumably antedated the genetic system that informs life as we know it.

## Growing (Inter)Dependency

Recall again that the final cause of something is simply its function or purpose, i.e., the *need* that it fulfills. So perhaps a good place to begin an inquiry into the teleological character of development is to ask: what engenders need? To ask that question we need to clarify the concept of 'need'.

A need can be defined quite generally as something that must occur in order for a configuration of mutually dependent and reinforcing processes, i.e., a *system*, to persist—an existential problem that demands a solution. For any complex system that persists by using free energy to maintain its 'self' against thermodynamic degradation, one obvious need is for free energy that can be used to support work: without it the system will disintegrate, and cease to exist. Thus to the extent that a system needs to persist, it has a need for free energy. This is why one system needs another—animals need plants, and plants need the sun. Indeed, were the sun to 'go out', most if not all life on earth would cease to exist, and the planet would rapidly progress toward thermodynamic equilibrium—that is, it would die.

The sticking point for science however is that while it is not a problem to rationalize what is needed for the persistence of a system, there is no *objective* reason why any particular system would *in and of itself* 'need' (and hence work) to find a solution to the problem of persistence. This conception of need is entirely subjective—based on an 'internal' sense of *selfness* that is refractory to objective ('externalist') discourse. Thus, to the question 'why does an organism work to persist?' science can only answer 'because it *inherited* the characteristic of persistence, mechanistically embodied in the genetic system (Dawkins's *The Selfish Gene*, 1976), from ancestors that successfully persisted (i.e., were hence naturally selected) after coming to be by lucky accident'. And maybe that is all there is to it.

But then again, maybe not—maybe the subjective or 'selfish' need to persist is not merely an illusion produced by replicative structures that are an accident of history, but rather bespeaks of the subjective nature of reality (Mathews, 2003). This alternative possibility, which evokes the image of a universe striving to express (and ultimately *know*) itself, cannot be empirically or logically refuted, and indeed, has many attractions—not the least of which is that it allows life and mind to be conceived as something other than an absurd accident. In any case, what you make of 'subjectivity'—whether you view it strictly as an emergent property of neurological sentience, or more generally as an intrinsic aspect of existence itself—basically comes down to what you choose to believe. It is important to note however that choices have developmental consequences that constrain the range of future possibilities, and that in this case it is not possible to prove or refute either proposition.

Be that as it may, it seems reasonable to assume that a system's subjective 'need' to persist is related to, and perhaps even stems from, its objective needs—in particular the need to consume free energy, which follows from the axiomatic proposition that complex systems always (eventually) respond to stress by adopting configurations that relieve it. Where might that lead us?

As we have already noted, this leads directly to the view that the Second Law of thermodynamics—the universal 'need', as it were, to equilibrate, i.e., relax, and thus release 'pent up' (potential) energy—is the final cause of everything (Salthe, 1993; Schneider & Kay, 1994; Schneider & Sagan, 2005). Development might then be viewed as a stress response entrained by the Second Law (Ulanowicz, 2009b).

This simple proposition allows us to tell the following tale of life's development:

The release of potential energy (i.e., equilibration) can either be destructive or constructive, depending on its rate and the material context in which it occurs. Fire is destructive, as is nuclear fission. But both can be materially *harnessed* for constructive *purposes*, as for example in internal combustion engines and nuclear reactors. Life on earth harnesses the fire of the sun as well as that which is vented from the interior of the earth, in order to construct itself. Such harnessing occurs in the metabolism of living systems: potential energy freed by chemical reactions is used by other chemical reactions for constructive purposes, including both repair and growth of the metabolic system that serves to release potential energy.

Growth can be defined generally as an increase in mass, structure, or productivity associated with a release of energy stored within a given locale. At the chemical level it is engendered by positive feedback cycles of metabolic reactions—for example the tri-carboxylic acid (TCA) cycle that forms the core of aerobic (oxidative) metabolism, or its evolutionary antecedent, the reductive tri-carboxylic acid cycle that Harold Morowitz and colleagues have shown was probably favored by pre-biotic conditions on primordial earth (Smith &

Morowitz, 2004). Such cycles increase the capacity for metabolism, in part through the production, as byproducts, of energy-carrying *currency* in the form of adenosine tri-phosphate (ATP), as well as efficient 'devices'—catalysts, structural scaffolds, and limiting membranes—that facilitate the metabolic reactions. Other devices might be produced that serve to repair and/or replace those that are lost to wear and tear. And so (as the story goes), by virtue of growing networks of interdependency, a self-referential, self-entailing, and self-reinforcing system of metabolism and repair is born.

Similar phenomenology, described quantitatively by theoretical ecologist Robert Ulanowicz (1997), characterizes the growth of ecosystems and economies. Ecology is all about positive feedback ('autocatalytic') cycles: the sunlight-fueled growth of plants fuels the growth of animal populations whose waste supports the growth of the plants. Similarly, in economies producers support consumers whose consumption generates currency and builds infrastructure that sustains the work of the producers. Positive feedback cycles at any level of scale engender the growth of whatever material energy-consuming system it is that embodies the cycle.

But how do such cycles emerge in nature? Certainly growth is a big part of the story. Positive feedback causes growth, amplifying the incipient asymmetry of the cycle itself by marshalling resources from the surrounding environment—a definitive aspect of autocatalytic cycles that Ulanowicz (1997) refers to as 'centripitality'. Such cycles are immanent under favorable conditions, owing to the combinatorially *immense* number of configurations that are possible in this universe (a quantitative aspect of complexity with profound implications; Elsasser, 1972; Ulanowicz, 2009a). Thus, systems of autocatalytic cycles might be expected to emerge spontaneously whenever circumstances afford sufficient combinatorial complexity, energy resources, and freedom from constraint.

Growth dominates the early or 'immature' stages of an emergent system's development, producing a burgeoning of creative *potential* (Salthe, 2003) that manifests as exploding diversity of system constituents and processes. In ontogeny one progenitor cell becomes a multitude of interacting cells. In ecological succession a few colonizing individuals explode into populations of interacting individuals. In human societies villages and towns harboring abundant economic opportunity (often by virtue of geography) grow into cities.

Immature systems are relatively chaotic in that, compared to mature systems, they behave erratically; that is, their specific behavior is highly contingent, and not (yet) predictable or 'settled'. Thus, for most organisms the mortality rate is high in early development and decreases with maturation, before climbing again in senescence, giving a U-shaped curve whose descending slope has been referred to as 'ontogenescence' (Levitis, 2011). Likewise, during brain development the rate of neural cell death is relatively high in early development and decreases during maturation (before climbing again in senescence). At the sub-cellular level, in animal ontogeny the number of different genes that are expressed (transcribed and translated into protein) decreases progressively within differentiating cell lineages. In ecological succession, species diversity declines as the ecosystem matures. The Cambrian Explosion of complex multi-cellular life produced a burgeoning of diverse body plans that were subsequently winnowed down to the relatively fewer represented in contemporary life forms (Gould, 1989). And in technology development toward a specific practical goal, the diversity of approaches or inventions tends to be high early (with a high rate of 'mortality') and later to settle on a few successful design variants—as for example with the automobile, or airplane.

What accounts for this developmental pattern of 'settling' into maturity, the emergence of a definitive or habitual way-of-being? While the self-(positive feedback)-driven growth of an system distinguishes it from background noise, the system can only grow so far before becoming stressed by resource limitations, as well as new frictional stresses produced by growth itself ('growth pains')—for example, inefficiencies caused by interference between competing or 'cross-purpose' processes that contribute to growth. If (as we assume) complex systems are compelled to adopt configurations that relieve stress, then something else is needed to relieve the stress of growth to these limits. This is where self-organization comes to play.

When growth of a cycle (or system of cycles) pushes inherent or imposed limits, the cycle itself becomes *selective* for those processes (often themselves cycles) that work best to feed the cycle (Ulanowicz, 1997). In ecology this manifests as the emergence of dominant species. In economics it manifests by economies becoming 'locked-in' to the adoption of specific technologies (Arthur, 1989). Inefficient or ineffective redundancies are eliminated through competitive winnowing or active repression, increasing functional precision. The result is a loss of indeterminacy, i.e., increased *information* (Coffman, 2011) that both constrains and creates potential along a more narrowly defined path of change. Differentiated specialization of processes is favored, leading to a division of labor that adapts a cycle to a more specialized resource 'niche' afforded by the activities of other cycles. At a higher level, multiple differentiated cycles become more effective at working together to release energy pent-up in less accessible locales. This process of self-organization culminates in the emergence of a mature (eco)system.

Natural selection, as envisioned by Darwin, is simply a matter of differential rates of organismal survival, and hence reproduction, driven by competition for limited resources. Whether an organism survives and reproduces depends to a large extent on how well it can take advantage of its circumstances to achieve those ends. But owing to positive feedback, *species* that are successful will continue to enjoy long-term success only insofar as their successes 'feed' the (eco)system, and hence (in karmic return) themselves. Natural selection can thus be seen as an aspect of *ecological* selection, the nature of which changes with development of (eco)systemic context. This is manifested in ecological succession: during its initial growth phase a newborn ecosystem favors rapid organismal reproduction ('r-selection'), whereas after the system has matured it favors niche specialization ('K-selection'). In human economic development the same phenomenology obtains, as systems develop from an immature stage favoring 'jack-of-all-trade' generalists to a mature stage favoring specialists. The self-organizing transition into maturity, wherein specialization becomes mutually beneficial, serves to relieve the stress of competition.

During the growth and maturation of life the Modeling Relation becomes a critically important means to the same ultimate end of stress relief. The Modeling Relation is crucial for understanding final cause, the basis for subjective knowledge. If a system is driven by need to persist and hence to consume free energy, it *pays* to anticipate, to prepare for opportunities and existential crises before they arise. Those systems that become able to successfully model their world will be more successful, more 'fit', than those that are not, and will hence be selected. Selection favors the creative ability to generate dynamic models of a world that is continuously changing the configurations through which it consumes (and within which it stores) free energy. As we will see, continued development of a specific system or configuration of systems builds up stress (stored energy)—and hence opportune potential—that can be relieved whenever some other system or configuration develops a model for its release. Hence, development produces conditions that 'call for' modeling, which is facilitated by (and occurs whenever configurations arise that fulfill) the Modeling Relation. Likewise, an ecosystem, via developed niches, 'calls for' specialized organisms to fill those niches.

But how do systems develop models of their world? This is a key question for understanding life itself. A fundamental part of the answer is implied by the term 'Modeling Relation', which indicates that a model is something that is *relational*. What this means is that modeling is an activity or process that occurs in one thing in relation to something else. Clearly from the perspective developed here both the thing that does the modeling (the subject) and the thing that is being modeled (the object) are complex systems, which means that they are composed of, yet cannot be fully understood in terms of, sub-systems, each of which also develops into existence, by way of some combination of growth, self-organization, and the Modeling Relation.

## Dialogic

At this juncture it behooves us to explore what it means to exist *in relation to* something, as the idea of relationality is the key to understanding how the Modeling Relation engenders development of subjective need. We normally think of 'being' as implying strict materiality; indeed, this is the basic presupposition of metaphysical Materialism. But if you reflect on the opening to Hamlet's famous soliloquy—"To be or not to be"—you will find that what Hamlet is referring to has nothing to do with material existence *per se*. For if Hamlet were to choose "not to be", that is, if he were to commit suicide, there would be no loss of his materiality at the moment of his death. And yet, Hamlet would cease to be. This brings to light the paradox of life enforced by the materialist perspective of science: that "to be" alive requires, in addition to materiality, something else, which is itself immaterial. We submit that this *vital* quality is a special kind of relationality; it is this which is lost at the moment of death.

What does it mean to exist 'in relation to'? And just as importantly, what sorts of things can something exist in relation to?

The concept of relationality implies asymmetry (and hence development), and is thus minimally dualistic: the existence of one thing 'in relation to' another means that the existence of one thing depends on, implies, or is in some way comparable to that of a second thing, which may or may not be present. The first two senses of the term ('depends on' and 'implies') are the most pertinent to the discussion here, as they provide a platform for seeing how existing 'in relation to' is but a step away from acting 'in order to'. If a process or system is intrinsically lacking something that it needs to persist, it can be said to exist (i.e., persist) *in relation to* (the need for) that thing. If the system persists, it does so because its relational need is satisfied by its actions. The probability that a system will persist is thus dependent on the extent that it acts in an informed way *in order to* satisfy its need. This is where the positive feedback loop of the Modeling Relation comes to play.

Recall that the act of modeling is fundamentally a creative process, wherein developed conditions external to a system present facts that can potentially be encoded as signs that can be processed (interpreted) by a developed entailment structure internal to the system, and thence creatively decoded by the system into action and form. What ultimately determines whether or not this process affords an anticipatory model is congruence between the act of decoding and actual (factual) environmental conditions—whether the decoded entailment *functions* to satisfy the system's needs. While the encoding, internal entailments, decoding, and external conditions are all materially embodied and can be understood in terms of whatever mechanistic determinants they happen to manifest, their congruent relationship is mediated by arbitrary signs and symbols, and hence, like language, music, and poetry, is indeterminate with respect to those mechanisms. The congruence is *created*, not determined.

The Modeling Relation thus describes what must happen for a system to enter into creative dialog with its environment in order to satisfy its needs. Anticipatory affordances—and thus ultimately, all practical knowledge—are a dialogical aspect of reality that take advantage of formal congruence between what are otherwise completely dissimilar phenomena (i.e., 'external' causation and 'internal' entailment), something that might be described metaphorically as 'resonance'. This is why we are able to describe the world using both metaphors and math.

It is easy to find examples of this anticipatory loop at play in the animal world. As I sit here typing these words my German Shepard is lying patiently on the floor, watching me. But as soon as I close my computer she will jump up and begin doing the "let's go" dance. The same thing happens when I finish breakfast and fold the newspaper. For her, my putting the computer down or folding the newspaper is a *sign* that I am ready to go. I may not actually be ready, but often enough I am, and her anticipatory goal of going for a walk is realized. And to the extent that her "let's go" dance motivates me to get going, it dialogically contributes to producing her 'intended' outcome, which satisfies the need that both of us have for exercise.

Somewhat paradoxically, anticipation (projecting forward) is a habitual (backward projecting) phenomenon. This too is illustrated by my dog's behavior. When we go for a walk she likes to play fetch, usually with a stick that she selects (in a way that is to my eye clearly not random). I throw the stick, and she retrieves it and brings it to me to throw again. Recently I changed the game to ball, which I throw using a launcher—essentially a sling that adds leverage to my throw. Normally when I throw a stick (or a ball with just my arm) she runs directly to where it lands. But when I started using the launcher I noticed that she would run only part way, then stop and look around, unable to find the ball, which had traveled twice the distance that I throw with my arm alone. Clearly she was anticipating where the ball would land based on her past experience with my usual (un-leveraged) throws, but her developed (i.e., habitual) anticipatory model was no longer congruent with the new reality. Only after a good deal of 'trial and error' searching was she able to develop a new model that allowed her to anticipate the distance the ball would travel when slung with the launcher.

We can learn a lot about relationality by interacting with (or better yet, *relating to*) animals, or even plants for that matter, and observing those interactions. The inter-subjective dialog itself becomes manifest, and reveals itself to be more relevant to life than the mere material facts of existence. A key aspect of this dialog is *signification*, the habitual specification of meaning—the interpretive use, by living creatures, of things and actions as symbols signifying other things and actions, which may or may not be present. This aspect of life, termed semiotics (or *semiosis*, referring to the process itself) was logically articulated by the American philosopher Charles Sanders Peirce (1839-1914), whose ideas anticipated those of Rosen and the thesis of this essay in many ways.

Peirce held that semiosis is minimally triadic, as it entails (1) an objective occurrence, (2) a sign of the occurrence consisting of *one* of its myriad effects, and (3) the subjective interpretation of that sign by an interpretant (or 'system of interpretance'; Salthe, 1993, 2009). This is essentially the first three steps of Rosen's Modeling Relation. But Peirce also anticipated the fourth step (decoding), because he noted that the process of interpretation creates signs that are themselves semiotically interpretable—either recursively by the system itself, or by other systems in the world at large. Indeed, the triadic structure developed by Peirce is formally similar to that of the recursive metabolism-repair relation developed by Rosen, which is quite remarkable given that Rosen does not appear to have been directly influenced by Peirce (Fernandez, 2008).

While it is easy to find examples of anticipatory semiosis in living systems of the contemporary world, the bigger question that must be asked is: under what circumstances does this process emerge in the evolutionary development of life? This question is beyond the scope of the present essay; here we will only sketch out some ideas pertinent to its framing.

Complex systems—especially immature systems, such as can be surmised existed on primordial earth—are inherently creative (some would say 'random' or 'stochastic'), and their creations (i.e., decoded entailments) need not have anything to do with the objective reality that exists outside the systems themselves. Such non-congruence does not mean that the system's interpretation is not real; it only means only that it does not *relate* (is irrelevant) to what is actually happening outside of the system, or to its needs—which is to say that it is not realistic in relation to the world. Needless to say, systems that are overly reliant on such decodings can be said to be unfit, and generally do not persist without an external means of support (a protective nurturing niche or 'asylum').

Thus, a *subject* can be defined as any complex system that can potentially develop an interpretation of the world. Since the world is itself a complex system, it makes sense to speak of the subjective nature of reality, and thus acknowledge that the latter is the dialogical creation of myriad subjects whose interpretations have some evaluable status *in relation to* the actual (factual) world at any given moment (Coffman, 2009). By the Modeling Relation, the internally entailed reality of a subject is a model to the extent that it is realized in the world at large, by way of functional form. But insofar as the system's decoded entailment *is not* but *yet might be* so realized, it affords *potential* for development toward that end. Thus, a subject can exist in relation to some need that might not at present exist, but which might yet come to exist in the future; and by working creatively to fulfill that need before it arises, to act 'intentionally'—even if completely unconsciously. Gould's ideas about evolutionary co-option ('exaptation') of pre-adapted structural motifs ('spandrels') fit this phenomenology (Gould, 2002).

As discussed in the previous chapter, the Modeling Relation represents a closed loop affording potential for positive (self-reinforcing) feedback. Since positive feedback engenders growth and self-organization, realization of a model facilitates development of a system. But such a system was engendered by, and exists within and in relation to, a larger or more general system that was *predisposed* to create it. For example, human consciousness (and with it intentionality as we know it) was engendered by neurologically embodied animal sentience, which manifests a less self-aware antecedent of intentionality. Similarly, animal sentience can be viewed as a developed form of subjective experience.

To frame this logically we can make use of the specification hierarchy, which models development (Salthe, 1993). As noted in the last chapter this hierarchy is a semantic model based on the logical dichotomy {*Generic*{*Specific*}}, which can also be expressed as {*Vague*{*Definite*}}, {*Implicit*{*Explicit*}} or {*Potential*{*Actual*}}. The dichotomy maps temporally to development as {*(Incipience)*⟶{*(Maintenance)*}} or {*(Initiation)*⟶{*(Positive Feedback)*}} (where the parentheses indicate the developmental condition or process inhering at each level of the dichotomy, and the arrows denote the trajectory of developmental change). In the specification hierarchy the generic levels are pre-requisites that are logically implied (i.e., entailed) by the specific levels that they include, whereas the latter are engendered (but not entailed) by vague predispositions of the former. Each included (i.e., more specific) level in the specification hierarchy emerges developmentally via positive feedback, which in the development of mind involves the Modeling Relation.

So, the developmental logic of consciousness can be represented as {experience{sentience{consciousness}}}. This hierarchy illuminates why we often rationalize our actions after the fact, without being consciously aware that that is what we are doing: sentient actions motivated by unconscious intent (e.g., by instinctive models based on genetic and/or neurological encodings and entailments) can reflexively elicit conscious rationalizations involving more highly developed linguistic models. Of course, the conscious rationalizations need not have anything to do with the unconscious intent motivating the action.

In this light the hemispheric differentiation of the human brain (McGilchrist, 2009) can be seen to constitute a developmental 'engine' for generating subjectively individuated consciousness, wherein the left hemisphere uses linguistic modeling to interpret raw (phenomenal) experience presented it by the right, thus making the implicit explicit, the vague definite, the metaphorical literal. Indeed, McGilchrist's thesis is that development of subjective experience is attentionally initiated by the right hemisphere and routinized by the left, which maps to the specification hierarchy as {(Right hemisphere) ⟶ {(Left hemisphere)}}. It thus makes sense that historically, in the evolutionary development of humanity, the psychosocial 'balance of power' has tended to progressively shift from right to left hemisphere dominance, as that is the natural developmental progression. According to McGilchrist (2009) this has occurred repeatedly throughout human history in a concatenated series of developmental cycles, each of which can be seen to begin with a cultural milieu favoring the widely vigilant mode of right hemisphere attention (e.g., the Renaissance) and end with one favoring the more focused mode of the left (e.g., the Reformation, or the latter part of the Enlightenment), which ultimately leads to societal collapse (e.g., the fall of Rome, or the French Revolution). Unfortunately, the scope of the collapse grows with each development, as the manipulative influence of left hemisphere is magnified—now to a global scale owing to the Industrial Revolution—by positive feedback, and enforced by repression of the right.

As should be coming clear, the development of anything has two possible resolutions: *metamorphosis* (a trajectory of implicit ⟶ explicit ⟶ implicit) or *extinction* (implicit ⟶ explicit ⟶ extinct). Although both types of progression occur, the latter is more common, and perhaps in our case more likely. On the other hand, the ultimate resolution of development—either by metamorphosis into a new implicit (immature) trajectory (a system adopting a new way of being), or by extinction—is inherently unpredictable, and therein rests our hope for the future. Of course, this is not something that the reductionist (left hemisphere) finds easy to 'grasp'.

## Subjective Individuation

To illustrate how relationality differentiates semiotic from non-semiotic developmental systems, we can compare an organism to an ocean wave. Both organism and wave are dissipative structures that develop into existence via growth and self-organization that relieve ('dissipate') the stress of potential energy by providing specific conduits for flow. Thus, in both cases flow engenders (and is implied by) form, giving {flow{form}}. Through such development, both an organism and a wave reach a state of 'self-organized criticality', in which there is little capacity for further development and the system is poised for collapse. This is seen as senescence of the organism, cresting of the ocean wave.

But unlike an organism, a wave cannot reasonably be said to act intentionally in relation to anything. While it is predisposed to act in a certain way by virtue of its formal circumstances, this does not (as far as we know) involve semiosis. Traversing the 'epistemic cut' (Pattee, 2001) that separates abiotic dissipative systems from life itself requires, in addition to growth and self-organization, a systemic embodiment of the Modeling Relation. Just as self-organization allows dissipative structure to emerge from thermodynamics (flow), the Modeling Relation engenders emergence of semiosis from structural dynamics (form), giving the specification hierarchy {flow{form{function}}}.

The common thread running through this hierarchy is relationality, which involves asymmetric juxtapositioning of at least two entities or systems (Deacon, 2011). Non-equilibrium thermodynamics (flow) is minimally relational, since it implies asymmetry in the form of disequilibrium. The development of structure (form) in turn requires interaction of (at least) two specifically different and asymmetrically opposed thermodynamic systems—e.g., a liquid hydrosphere (ocean) and gaseous atmosphere—a minimal relationship through which dynamically emerges a more specified level (e.g., ocean wave). Semiosis (function) then requires asymmetric interaction of two or more formally structured systems—e.g., those embodied by the genome and cytoarchitecture of a cell—via the Modeling Relation.

The clarity afforded by the developmental logic of the specification hierarchy can be appreciated by considering the age-old conundrum: what comes first, the chicken or the egg? Given the hierarchy {amniote{bird{chicken}}}, we can confidently answer: the egg comes first. Before there were chickens, the ancestors of chickens laid eggs, and one of these developed into the first chicken. Because the concept of 'egg' is more generic than 'chicken', the egg *must* come first in the evolutionary development of life. (On the other hand, if we were to ask 'what comes first, the chicken or the chicken egg?' the answer is, by the same logic, 'the chicken'.)

Development through increasing levels of specification entails a progressive acquisition of informational constraint, which is also relevant to the problem of individuality. In the evolution of animal life, it is reasonable to ask when animal individuality first emerged (Buss, 1988), just as it is reasonable to ask when an individual human being first emerges in ontogeny. Neither question provides a clear-cut answer unless *we* define individuality. If we accept the premise of this chapter that any specifically definable thing comes into existence developmentally, then if follows that the same is true for any individual. An individual is a specifically embodied example of a broader class of things, and its/his/her development can thus be modeled using the specification hierarchy, e.g., as in {human being{Abraham Lincoln}}. But what distinguishes an individual from any other individual of the same class is not just development, but evolution—the contingent spontaneous change that occurs within the constraints erected by development, some of which are unique and irreproducible. Individuation can thus be viewed as the evolutionary change that befalls a developing system (Salthe, 1993).

So, in the development of life on earth, subjective individuality *as we know it* has become more specifically 'focused' in some animal forms than others. An individual sponge or a hydra is not an individual *subject* in the same sense as a frog or human being, because their relational character is much less constrained at the organismal level. A sponge can be dissociated into single cells, and a new (and somewhat different albeit genetically the same) sponge developmentally reconstituted from those same cells. A hydra can be cut into pieces, each of which will developmentally regenerate a new hydra that is genetically identical to the first. Neither sponge nor hydra has a brain. And while every sponge and every hydra is individuated by virtue of its unique history, this is an objective attribute that differs from the subjective individuation that we recognize as personality in human beings and other so-called 'higher' animals.

In contrast to sponges and hydras, individuated frogs and human beings do not have the capacity to regenerate new frogs and new human beings after being cut into pieces, and have brains (not to mention hearts and other specialized organs) that are essential for their individual survival. To the extent that their brains enable anticipatory behavior, frogs and human beings are more highly developed intentional agents than are sponges and hydras, with a more refined ability to pay attention to and interpret environmental conditions that signify information that is potentially relevant to their persistence. Unlike sponges and hydras, frogs and human beings are recognizable as sentient individuals. This is a relational characteristic that emerges via the more anatomically sophisticated 'division of labor' that develops in the vertebrate animals.

But subjective intentionality as we know it is far more developed in human beings than it is in frogs (or even dogs or chimpanzees), being not merely sentient, but conscious. Signified information is not only attended to and acted on (something that all life forms do to some extent), but articulated and processed linguistically and metaphorically. And communication via this language is not only mediated by acoustic wave-forms, but extends to writing and electronics. By virtue of this higher level of semiosis, human beings are not just aware, but aware of being aware. With (self)-consciousness, normative values become encoded linguistically into morals. And with this uniquely human level of awareness, innocence is lost.

Or is it?

Human individuality also emerges progressively with the development of each individual human being. It is *not* established at conception, for by chance a single conceptus can and often does develop into multiple individuals (twins, triplets, etc.) that are genetically identical. Moreover, multiple embryos can and sometimes do fuse to form single individuals that are hybrids of genetically distinct concepti. Sentient individuality does not emerge until sometimes after the brain develops. Self-aware human consciousness (as we know it) does not emerge until well after birth, by virtue of linguistic interactions with other human beings during the first few years of life of a developing individual. And (depending on how narrowly you define consciousness), it may never fully emerge in some individuals—as defined by Jaynes (1976), it is not well-developed in schizophrenics [or perhaps, as suggested by McGilchrist (2009) it is overdeveloped, to the point of collapse], just as it was not yet emergent in our 'bicameral-minded' ancestors as recently as ~3,000 years ago. In this view, even in modern humans conscious awareness is still relatively undeveloped in many people and highly variable across the global human population.

This brings us to an important and somewhat paradoxical point in regards to the concept of intentionality, and its relationship to human consciousness, morality and the notion of innocence. For in colloquial usage 'intentional action' implies conscious awareness of one's intent. And yet we have just argued that consciousness as we normally conceive it emerged relatively recently in human evolution, and is probably a highly variable cognitive trait even now among modern humans. To the extent that moral behavior requires conscious intent one can argue that many (perhaps most!) human beings are, like their animal cousins and pre-conscious human ancestors, essentially innocent because they are not fully conscious of what they are doing much of the time. In this light Jesus' words "they know not what they do" can be interpreted quite literally, and the legal defense "innocent by reason of insanity" may merit significantly broader application.

But this narrow definition of conscious intent does not mean that intentionality—that is, goal-directed (teleological) action—is uniquely human, as is generally assumed, because we can discern a specification hierarchy wherein the conscious manifestation of intentionality emerges via evolutionary development from a vaguer (less conscious) form of sentient intentionality, which in turn emerges from an even vaguer form of experiential intentionality, giving as noted above {experience{sentience{consciousness}}}. Thus, a sponge 'intends' to do whatever it does (grow, reproduce) only in the vague sense of using its experience, in the form of the environmental cues it receives via the rudimentary sensory apparatus with which it is endowed, to change its rate of pumping, to secrete defensive toxins, etc., which further the sponge's 'goal' of growth and reproduction. A similar level of unconscious and largely insentient 'intent' can be ascribed to plants, fungi, and unicellular organisms, although the circumstances and interpretive behaviors are specifically different. It therefore makes sense to wonder what it is like to be a sponge, an oak tree, a slime mold, or even a bacterium.

The more developed, highly centralized sensory system of a frog allows it both to specifically focus its intentions, and extend them over a wider domain. A frog is more acutely aware of (some specific aspects of) its world. By lying in wait at the edge of a pond, watching, a frog 'intends' to capture a passing fly; by singing, it 'intends' to attract a mate. But it does so unconsciously.

So, intentionality has developed, via the evolutionary ratchet of nested developmental cycles, through successively higher (more refined or specified) levels of experiential awareness leading to human consciousness. But we might well surmise that it has also developed, via evolution, to high levels of specification that are unique to each species. This is because the specification hierarchy produced by evolutionary development takes the form of a tree (Salthe, 2010), with higher (more specified) levels branching from a common trunk of lower (vaguer less specified) levels.  So while human beings, frogs, and sponges can all be said to subjectively experience their world, the way that this experience actually manifests in each form is specifically different.  Sponges and frogs have not developed consciousness, but they nonetheless have a unique way of relating to the world as developed subjects that we (as quite differently developed subjects) cannot possibly relate to.

By extension, even among human beings we might expect that consciousness in each individual is somewhat different—we all experience the world in uniquely individuated ways that are not captured by the generic concept of 'consciousness'.  So 'higher levels' of consciousness refers to more specified (and perhaps as yet undeveloped) veins of conscious experience, of which there are many, each of which is granted unique insights.  As individuated subjects we are each, to a very large extent, alone in the world.  And yet, since human consciousness is widely shared via highly complex languages involving myriad disparate artistic and technological media, its insights can (to some extent) be taught and learned.  And that is a big part of what sets humanity apart from the rest of the animal kingdom.

This is why it is easy to fall into the trap of believing that subjective agency is uniquely human: it both is and is not.  Human consciousness *is* unique, a highly developed mode of experience, and a game-changing emergence such as only rarely occurs in the evolution of life—a global transformation perhaps comparable in magnitude to that which followed the evolutionary development of photosynthesis two billion years ago.  And because of it we are special: we are granted knowledge (and hence power) that is not available to mere sentience.  But that does not mean that *Homo sapiens* are the only subjective agents on this planet.  It just means that we are the only agents in which intentionality has developed enough to be wielded intentionally.

### Entrenchment

Returning now to our premise that complex systems respond to stress by eventually (and often creatively) adopting configurations that relieve it, we can view development as a trajectory of change through which growth, self-organization, and the Modeling Relation are used in various ways and to varying degrees by hierarchically nested systems toward the same ultimate end of allowing energy to flow more freely. Which is all well and good, but it has its downside for the systems themselves: increasing dependency on, or need for, some specific thing. First, growth is based on a system's mobilization of a specific energy resource, which means that the existence of the system that is doing the mobilizing is dependent on the resource. Second, self-organization is based on a division of labor through differentiation and specialization, which means that the constituents of the developing system become more and more dependent on the overall system, and on each other. Third, successful implementation of the Modeling Relation requires congruence between the model and the thing being modeled, so the former depends on the latter not changing in a way that degrades or destroys that congruence.

All of these developed dependencies can be seen in the disruptive effects of climate change on the seasonal dynamics ('phenology') of trophic interactions among species in temperate latitudes (Durant *et al.*, 2007). For example, many bird populations migrate to northern breeding grounds in the spring, in order to feed their newly hatched chicks on insects that emerge during a short window of time corresponding to the spring growth of plants. This illustrates the dependency of growth—of the bird, insect, and plant populations—on resource availability, which is seasonal. Some species of bird are highly specialized to feed on a specific species of caterpillar, which is in turn specialized to feed on a specific species of flower. This illustrates how organization—in this case of an ecosystem—increases the interdependency of system constituents. Finally, it is sometimes the case that the phenology of the bird migration is cued by signals unrelated to climate (e.g., day length), whereas the phenology of their prey is cued by temperature. The mismatch between the phenology of the predator and that of the prey can lead to birds arriving too late to their breeding grounds, after their caterpillar prey have developed into moths and are thus no longer present. The result is starvation and decimation of the bird population. This illustrates the dependency of a model on congruence—the migratory phenology of the birds embodies a model of their world only as long as it remains synchronized with the ontogenetic phenology of their insect prey.

Development selectively reinforces specific dependencies while eliminating alternatives and/or redundancies, which tend to impede energy flow along a favored path (while enabling it along alternate paths). If and when a resource becomes limiting, the specialists who are more adept at accessing that resource tend to out-compete generalists, and the diversity of components and flows within the system decreases. A system thus becomes more organized or mutually 'informed'. But generalists and disorganized diversity afford degrees of freedom. So the more developed a system is, the less potential (higher resistance) it has to change its *modus operandi* to access an alternate path, particularly when energy becomes limiting. The capacity for development of any specified system is limited by the developed constraints that specify the system, and yet such capacity is required for a system to change and hence to adapt. Hence, in evolution most species become extinct (owing to overspecialization and consequent loss of adaptive plasticity), and ecological collapses occur regularly.

So development is somewhat predictable, but this predictability differs fundamentally from that of mechanisms. Whereas mechanisms are determinate, development involves—and in fact depends on—*in*determinacy. What this means is that unlike mechanisms, development is predictable only in a statistical sense: that is, in terms of probabilities. Moreover, the distribution of these probabilities is dynamic: mature systems are more predictable than immature systems because they are more determined, which simply means that their internal configurations and overall behavior are more constrained and hence less 'random'. This is different from the predictability of a mechanism. In fact, from this perspective we can see that the rigid determinacy of mechanisms represents the *limit* of development: a given mechanism is fully determined, and hence a fully developed entity: the ultimate realization of a model. Its refinement can only occur through creative intervention, which involves some *deconstruction* that introduces indeterminacy (uncertainty) and hence unpredictability. As such, true mechanisms only exist within, and as part of, developing systems that retain the capacity to purposefully create and repair them—for example, human systems.

In other words, mechanisms are not the *driving* cause of anything; they are merely the developed means to an end.

Here it is useful to reprise Rosen's answer to the question of how life differs from a mechanism. The essential difference is that a mechanism, as objectively conceived within the Baconian-Cartesian-Newtonian framework, works simply by virtue of dynamically changing external forces, i.e., efficient cause from the outside. So within this framework no mechanism can be explained without invoking another mechanism, and so on *ad infinitum*. The only way to truncate this infinite causal regress is to invoke a causative agent that is external to nature (i.e., a *super*natural agent)—the classic cosmological argument for a deity. The mechanistic framework of reductionist science thus reinforces the rational cognitive 'need' for the supernatural.

Life, in contrast, is causally complex, and can be reasonably viewed as a teleological response to a *need* that arises *within* a system. A compulsion to satisfy that need (and thus reduce stress) is what causes a system to develop. Toward that end it can grow, self-organize, and create embodied models—mechanisms such as the genetic system—that help it grow and self-organize. Developed configurations, including formal (virtual) configurations embodied as models, are themselves causal in that they constrain potential (Juarrero, 1999). They persist as long as they are useful, and are ultimately disposed of when they get in the way of satisfying the needs of the larger system within which they developed.

The logic of development is thus recursive: it is about systems of interacting autocatalytic cycles—including the world at large—using whatever is materially, efficiently, and formally available to develop ever-more *refined* ways of satisfying stress-inducing needs that originate within the system, and which are hence, from the perspective of the system, entirely subjective.

So, to summarize the argument developed thus far: the development of any specific thing or configuration is a creative system-level response to the stress of potential energy or interference ('friction') between different processes, i.e., *resistance to flow* (Bejan & Lorente, 2010). Development relieves such stress through a combination of (1) growth, a 'constructive' conduit for rapidly dispersing potential energy within a given locale; (2) organization, which relieves frictional interference by entraining cooperation among progressively differentiated specialists; and (3) modeling, which enhances growth and organization by way of anticipation. As a result of development, things become more predictable—more mechanistic. But the negative consequence of this is over-commitment to (dependency on) a narrow range of resources and/or models: specialization reduces the plasticity needed to relieve stress. And in systems that are far from thermodynamic equilibrium, stress is always present in ever changing ways, so development is unsustainable and ultimately entails death or extinction of the developing system—unless the developmental process can somehow be 'reversed' or 'arrested', or used as a platform to launch a new developmental trajectory with a somewhat different set of dependencies ('metamorphosis').

Maintaining oneself in prolonged or 'metastable' state of maturity—a state that is developed enough to work effectively, but which retains enough plasticity to sustain homeostasis and regenerative capacity (i.e., *health*) in the face of a changing environment—is a balancing act. It is exemplified in some animal species whose adult forms have 'negligible senescence' (Finch, 1990, 2009): for example, red sea urchins (*Strongylocentrotus franciscanus*) can live for over 150 years with no signs of decrepitude, including no loss of fecundity (Ebert, 2008). Interestingly, most if not all animals with negligible senescence have indeterminate growth—they keep growing throughout their lifespan. Prolonged maturity in this context involves continued growth within a mature structural context that allows developed constraints and damage (also a constraint!) to be continuously jettisoned. It's a treadmill, as it were: a cycle of suspended animation or *regenerative homeostasis*.

On one side of this treadmill, damage is removed through programmed cell death (apoptosis): the damaged cells simply kill themselves. On the other side (in some animals at least), signals released by the dying cells elicit new growth from quiescent multi-potential (stem) cells (Bergmann & Steller, 2010). Indeed, stem cells (including 'germ cells' that give rise to gametes) are another case in point (at a lower level), as they themselves undergo asymmetric cell division in which one of the two daughter cells becomes a proliferating progenitor cell mounting a developmental response, and the other remains an uncommitted stem cell. We can speculate that during stem cell division accumulated damage is asymmetrically shunted into the daughter cell whose lineage will go on to differentiate, allowing the stem cell lineage to remain relatively undamaged (and relatively uncommitted!). And indeed this is what happens in budding yeast: accumulated damage is selectively retained in the 'mother' cell, which thus undergoes aging over a limited lifespan that leads to its eventual inability to produce any more daughters ('replicative senescence').

Mature ecosystems—a rainforest for example—manifest an analogous treadmill of life coupled to death. Death serves the purpose of freeing up resources, providing nourishment for new growth, and new living space. To persist, life *needs* death.

Development can be suspended in maturity only if it is coupled to some form of deconstruction that allows for new growth. Otherwise continued development causes the system to become over-committed, and as a consequence of the strict dependencies thus enforced, inflexibly predictable, like a machine. Ultimately continued development of the system itself becomes all but impossible, producing a situation with a predictable, if not always desirable, outcome: collapse of the system.

Regenerative repair requires spare capacity for development, so organisms that develop (either ontogenetically or phylogenetically) to the limit of their capacity lose the ability to repair themselves and become senescent, susceptible to the depredations of the Second Law of thermodynamics. In biological evolution this is likely to relate in some way to how far development has progressed in each lineage, which depends on ecological context and the specific energetic tradeoffs that occurred to meet the cost of living during the history of that lineage. Similar phenomenology obtains at the level of ecosystems, which also undergo metabolism and repair in a developmental context of growth and regeneration. As Ulanowicz (1997, 2009a) has shown, the health and resilience of an ecosystem or economy depends on how much uncommitted 'overhead'—measured as statistical indeterminacy and manifesting as diversity, redundancy and unpredictability—it sustains. Diversity and redundancy are what afford regenerative capacity. So when they are developmentally sacrificed (e.g., for the sake of machine-like efficiency and predictability), the system loses the ability to repair and regenerate itself following major (and ultimately even minor) perturbations.

And as we shall see in what follows, this is what has happened to contemporary civilization via development of the Global Economy.

# Chapter 5

# METABOLISM AND REPAIR IN THE GLOBAL ECONOMY

At this juncture it is worth distilling some key insights and implications from what came before in order to shed as much light as possible on what follows. We have argued that living systems (including human beings) create knowledge by developing models, a dialogical process manifesting the Modeling Relation. The development of anything (including of models) entails growth and selective self-organization of a materially embodied system, which produces metastable constraints on thermodynamic flow by probabilistically decreasing resistance along some paths of potential while increasing resistance along others. The development of models occurs via positive and negative feedback engendered by the circuitous form of the Modeling Relation. A model emerges only when (and persists only as long as) the entailment structure of a system exists, via encoded signification and decoded action, in a congruent relationship of *fitness* with the real world. In this light, Darwinian evolution by natural selection can itself be viewed as a process of model development.

What this implies, as we hope should now be clear, is that development of a model requires a subject (i.e., a living system), as well as a realistic relationship between that subject and its world. That is, the logical entailments of a subject must, when decoded into action, be realized by some objective aspect of the actual world that in turn fulfills the persistent needs of the subject. For that to occur reliably, the entailment structure within the subject must have already developed a realistic relationship with some aspect of the external world, that is, some sort of encoded *memory*—be it within the structure of DNA, neural connections, cultural mythologies, or whatever—of how to go about developing a realistic model. Thus any system that embodies a model transcends the present moment by purposefully using its internal record of history to (try to) direct change toward its own need-based ends. This is the essence of teleology, the logic of final cause.

So in essence, through the Modeling Relation a living system anticipates the future by recalling the past. Because models must (like anything else) develop into existence, they are most effectively realized at maturity, which is the stage of maximal congruence between subject and object: immature models have a minimal record of history, and hence relatively few decoded entailments that are realized long-term, whereas senescent models are developmentally committed (locked in) to realized aspects of the past that are decreasingly relevant to the subject's future needs.

Intriguingly, neurological studies of the human brain suggest that consciousness is not a direct experience of external sensations, but rather of associative memories elicited by those sensations—that is, conscious perception occurs via recalled memory (Meyer, 2012). In other words, *we experience the present and anticipate the future by living in the past.* This might explain why change becomes increasingly difficult as we age. It also explains how a well-developed human socioeconomic system, such as the contemporary Global Economy and the mythology of mechanisms upon which it is founded, actually interferes with our ability to think differently in order to solve the problems created by that very system: the model that the system embodies has become encoded into the memorized entailment structure that directs both our present experience and anticipatory thoughts. Hence, just as individual organisms develop into senescence, so to do the higher-level systems they create as the embodied realization of their models.

Furthermore, one might expect that, given that development tends to reduce the undeveloped diversity upon which regenerative capacity depends (as discussed in the last chapter), continued development of a high-level system—a Global Economy—not only via the actions of a single species, but by way of a single narrowly focused model developed by that species, will eventually pose serious adaptive difficulties for that species as the model falls out of congruence with reality. And this is indeed the problem with which we are now faced.

Owing to the 'monoculture' of models realized by the mythology of mechanisms as embodied in modern technology, the human world is easily distinguished from the non-human world: our machines and other technological artifacts—cars, televisions, computers, telephones, electric lights, plastic containers, etc.—appear, on a backdrop of nature, strangely unnatural. Most of us are completely immersed in this manufactured world and depend on it for our survival, and as a result we too seem disconnected from nature. Many of us (particularly those held under the sway of literalist strains of religion) are quite convinced that we, unlike any other species of animal, are *above* nature.

But of course we are not. We are animals—clothed apes—and everything we do is as natural as anything else in this universe. Nevertheless, from the perspective of planet earth, we are also quite new and different—and as a species, exceptionally dominant. Our unique form of language-based consciousness grants us deep insight into the mechanical aspect of nature, allowing us to bend her to our will in ways that no other animal can. Arthur C. Clarke (1962) and Umberto Eco (2000) have noted that any sufficiently advanced technology is indistinguishable from magic. Through reductionist science, human beings have accessed, like Goethe's *Der Zauberlehrling* (The Sorcerer's Apprentice), some of the 'deep magic' of nature.

Despite the uniqueness of our gift for engineering, it is worth considering how the historical ascent of contemporary civilization fits the model of nature articulated in this essay. Western civilization is (part of) an 'autocatalytic' human socioeconomic system that begat and developed a powerful cognitive model—a mythology of mechanisms—that was put to use in developing the Global Economy. As with any development, the result was that civilization committed itself to a relatively narrow range of dependencies and options, upon which its persistence now depends. As we have argued, development of any system is inherently unsustainable. Sustainability—the ability to prolong maturity without entering senescence—requires a means of dynamically suspending (or reversing) development, at some level, in order to retain capacity for adaptive change. As discussed above, in organisms with negligible senescence this occurs through the balanced coupling of continuous growth with continuous *deconstruction* (e.g., via programmed cell death or de-differentiation). A similar phenomenology can also be seen in ecosystems.

Throughout history civilizations have developed and flourished for a time, only to suddenly collapse. In his book *Collapse* Jared Diamond (2004) provides an insightful answer to the question of why this happens, which fits well with the proposition that the rise and fall of societies is essentially a developmental phenomenon. Like bacteria in a Petri dish, growth of a civilization typically eats away at its resource base, and psycho-socio-cultural traits (cognitive models commonly referred to as 'traditional values') that contribute to a society's success during its growth phase become entrenched impediments to change when growth itself becomes detrimental, and the system becomes vulnerable to destabilizing 'externalities' (e.g., accumulating toxins in the environment and hostile neighbors). This developmental phenomenology is readily apparent in the historical emergence of the Global Economy, a complex system of 'metabolism and repair' that now holds humanity in its grip.

Consider again that any development is a teleological response to some need. The basic needs of human beings are the same as those of any other animal: food, water, and a means of protecting and defending one*self* against existential threats—the sinister 'other'. As we are social animals, we meet those needs by purposefully cooperating within groups—primitively, tribes defined by familial relationships, and later villages, kingdoms, city-states, etc., defined culturally by common stories and systems of belief, i.e., mythologies. The emergence of a society manifests development to the extent that it involves growth (of a population), organization (via division of labor), and the Modeling Relation (mythology and culture) that differentiates subjective self ('us') from objective other ('them').

But, despite our uniqueness, it bears repeating that everything we are and do, as individuals and as a civilization, extend from and reflect the fact that we are *animals*. So, our account of metabolism and repair in the Global Economy needs to begin with a discussion of what it means to be an animal, and how that way of being, with all of its inherent, inescapable *animal needs*, entrains our activities.

To be an animal is, first and foremost, to *need to eat*. It is our (i.e., the animal kingdom's) special (i.e., developed) way of expressing the universal 'need', discussed in Chapter 3, to degrade (i.e., release pent up) energy. It is the primary reason why we are *animated* by *appetite*—an antecedent of pleasure, not to mention greed and addiction. Accordingly, like all other animals (but unlike plants, whose sustenance flows directly from the sun) we are equipped with nervous systems that allow us to seek out, identify, and ingest food. The rest of our anatomy—musculature, digestive system, even skeletal systems—are also 'geared' toward the same end of eating food. To be sure, our anatomy also works to keep us alive by both distributing the nutrients throughout our bodies and detecting and avoiding or defending against threats, and also to reproduce—fundamental needs shared by all life. But the specific existential need to eat, in the strict sense of obtaining sustenance via the ingestion of food, is a defining (and hence developed) characteristic of animals not shared by other forms of multicellular life. As animals, only by eating can we maintain the complex physiological systems of metabolism and repair that keep us alive.

It makes sense then to say that the developed infrastructure mediating the economic 'metabolism and repair' that sustains civilization is based on, and represents a higher-level organized expression of, our animal need to eat. This is not a stretch—it is widely accepted that civilization grew out of the invention of agriculture, specifically the farming of grains in the Fertile Crescent of the Middle East. Agriculture for the first time allowed human beings to accumulate and thus *bank* food, which promoted the development of hierarchically stratified social classes defined by wealth—the 'haves' who control (own) the land and surplus food (and protect that ownership by paying for armed protectors, i.e., police and armies), and the 'have-nots' who work for them. The ostensible upside of this development was that it supposedly granted opportunities (especially to the 'haves') not available to hunter-gatherers, whose activities are more immediately constrained and entrained by the dynamic vagaries of their environment.

So with agriculture, a developed response to our animal need to eat, humanity built a economic system of 'metabolism' (growth, harvesting, processing, storage, distribution, and preparation of food, and construction of the infrastructure to do those things) and 'repair' (maintenance of the infrastructure) that granted at least some people a measure of increased autonomy, and that was also more insular—a system that, in short, allowed us to distance ourselves from the complexity of the world, and begin to impose our own patterns. Both cities (which epitomize human insularism) and written language (which Julian Jaynes argued catalyzed the emergence of human consciousness) codeveloped with agriculture. Ironically, one of its side-effects was malnutrition, a malady that affects the vast majority of the human population to this day (and not only in the 'undeveloped' Third World). This then is how we began to lose touch with reality. Diamond (1987) has thus argued that agriculture was "the worst mistake in the history of the human race".

That being the case, why did our ancestors adopt it? According to Diamond it was a response to the increased need for food caused by the growth of human populations, a hypothesis that actually follows from the developmental logic discussed in the last chapter. In other words, agriculture was a self-organizing (selected and selective) response to the energy stress of population growth, which broke symmetry by producing a division of labor and social stratification. As a result, the "elite became better off, but most people became worse off" (Diamond, 1987):

> As population densities of hunter-gatherers slowly rose at the end of the ice ages, bands had to choose between feeding more mouths by taking the first steps toward agriculture, or else finding ways to limit growth. Some bands chose the former solution, unable to anticipate the evils of farming, and seduced by the transient abundance they enjoyed until population growth caught up with increased food production. Such bands outbred and then drove off or killed the bands that chose to remain hunter-gatherers, because a hundred malnourished farmers can still outfight one healthy hunter. It's not that hunter-gatherers abandoned their life style, but that those sensible enough not to abandon it were forced out of all areas except the ones farmers didn't want.
>
> At this point it's instructive to recall the common complaint that archaeology is a luxury, concerned with the remote past, and offering no lessons for the present. Archaeologists studying the rise of farming have reconstructed a crucial stage at which we made the worst mistake in human history. Forced to choose between limiting population or trying to increase food production, we chose the latter and ended up with starvation, warfare, and tyranny.
>
> Hunter-gatherers practiced the most successful and longest-lasting life style in human history. In contrast, we're still struggling with the mess into which agriculture has tumbled us, and it's unclear whether we can solve it. Suppose that an archaeologist who had visited from outer space were trying to explain human history to his fellow spacelings. He might illustrate the results of his digs by a 24-hour clock on which one hour represents 100,000 years of real past time. If the history of the human race began at midnight, then we would now be almost at the end of our first day. We lived as hunter-gatherers for nearly the whole of that day, from midnight through dawn, noon, and sunset. Finally, at 11:54 p.m. we adopted agriculture. As our second midnight approaches, will the plight of famine-stricken peasants gradually spread to engulf us all? Or will we somehow achieve those seductive blessings that we imagine behind agriculture's glittering facade, and that have so far eluded us? (Diamond, 1987)

Agriculture was the initial stage of modern economic development, the globalization of which arguably began when human beings became capable of routinely crossing the ocean in ships large enough to transport goods. This enabled the conquest of as yet 'undeveloped' continents (still largely inhabited by hunter-gatherer clans) by the armies, mercenaries and missionaries of agricultural (and later industrial) kingdoms and empires. So it is reasonable to begin our story of the ascendancy of the modern Global Economy with the European conquest of America.

The conquest began with Columbus in 1492 and (for intents and purposes) ended with the Wounded Knee massacre in 1890. During that 400 year interval Native Americans were forcibly displaced and largely annihilated by Europeans and their descendents whose religious faith supposedly granted 'God-given' rights that the natives lacked. As Diamond has pointed out, the actual advantage the Europeans had over the Native Americans was neither spiritual nor intellectual, but simply a long history of *Guns, Germs, and Steel* (Diamond, 1997). In other words, owing to environmental (geographical and historical) contingencies, Europeans had developed both immunological and technological advantages, which granted them greater power in the form of economic leverage. They also had well-developed cognitive models—epitomized by the twin pillars of Religion and Science—that, through *ex post facto* rationalization enabled by objectification of the 'other', worked to repress the innate empathetic sense that instinctively inhibits inhumane behavior.

Of course, Western civilization has no monopoly on inhumane behavior. Violent acquisitiveness is an innate aspect of animal and hence human nature. One need only look at how different clans, both human and non-human, behave toward one another to see this. Few if any human societies have transcended inhumanity towards the culturally- (and genetically-) defined 'other'. We animals are driven by a need to consume, and when someone or something gets in our way, we strive mightily to eliminate that impediment, especially when that someone or something is not our close kin. But we are complex systems, so that is only one aspect among many in our nature. As conscious creatures we can supposedly choose which aspects to celebrate, and thus to nurture.

And yet, as our ability to destroy the 'other' grows, we naturally become more focused on, and entrained toward, developing even greater capacity to destroy. While this is generally viewed as a defensive behavior, it increases the probability of self-destruction, an ironic side effect of channeling resources into the development and deployment of defensive forces armed with destructive technology. The 'guns or butter' argument about allocating resources becomes a self-perpetuating aspect of politics (Reich, 2007; Roszak, 1972; Soros, 2006, 2008). While it is impossible to perform a controlled experiment to test the proposition, it appears that socio-economic-political systems selectively favor those who demand that we continue to arm ourselves with increasingly sophisticated and evermore deadly weapons. Meanwhile this same segment of the cultural and economic demography, the beneficiaries of the military-industrial complex that Eisenhower warned us about, becomes wealthy and politically powerful.

In any case, the point here is not that the European conquest of America (and everything that followed) was uniquely inhumane; it is that it was uniquely enabled, on an unprecedented scale, by science-based technology. The Enlightenment following upon the natural philosophy of Bacon, Descartes, and Newton led directly to the Industrial Revolution, which once again transformed the way human beings eked out a living. While this transformation occurred throughout the Western world during the nineteenth century, in the United States it came to a head in the War Between the States. With the victory of the Union, the agrarian aristocracy of the South, which benefited from and depended on the forced labor of African slaves, was forcibly supplanted by the more powerful industrial aristocracy of the North, which benefited from and depended on the less overtly racist (and hence cognitively more palatable) wage-based labor that engendered 'equal opportunity' enslavement to factories.

With the movement from agrarian to industrial economics, the 'metabolism' of human societies became decreasingly dependent on the labor of animals (both non-human and human), which is energized entirely by the sun-fueled ecology of extant life, and increasingly dependent on the far more powerful labor of machines, energized by the oily deposits of extinct life rendered by millions of years of geochemical activity. The energy released from those limited 'reserves' (precisely analogous to the yolk of an egg that fuels the early development of an embryo) fueled an unprecedented explosion of the human population—the growth phase in the development of the modern Global Economy.

But machine-powered labor is merely a means to an end. To answer the question 'to what end is the Global economy developing?' we need to identify its assorted and overarching purposes. Here the complexity of the system, and of the problem, becomes apparent.

Ostensibly, the overarching purpose of the Global Economy is to provide for people's needs. The economic 'metabolism' functions by extracting resources from the earth—minerals and petroleum, food and water—which it refines and transforms into goods, services, and accumulated wealth. To be sure, the distribution of the latter has always been extremely uneven, with a vanishingly small fraction of the population controlling the lion's share. Those engaged in actually producing the wealth (the laborers) have little opportunity to enjoy the fruits of their labor, which goes instead to those who *legally own* (and thus afford) most of the land, goods, services, and/or capital, be they members of capitalist corporations or socialist (or 'communist') states.

But this is the same as it ever was, at least in post-agricultural societies. The economies that we label 'capitalist', 'socialist', and 'communist' are really just different political systems for dispensing and protecting legal ownership (i.e., economic privilege). Since human beings tend to cede power to bullies and sociopaths, our economic systems have very little to do with the ideals of Adam Smith and Karl Marx (Eagleton, 2011; Magdoff & Foster, 2011). Likewise, the reason that the United States of America is a 'free country' has little to do with the myth of equal opportunity for all, and everything to do with economic *privilege* gained through plundering of natural resources—land and water taken forcibly from the Natives who lived here before the European's arrived, and from future generations. Much of the wealth and power was initially amassed through the use of slaves; but even after slavery became less useful and was finally outlawed, exploitation of humans continued, legally protected by a stacked deck of ownership laws enacted by politicians beholden to the rich and powerful.

Nevertheless, something certainly appears to be working to the benefit of many, at least in the developed world, because more people are living in relative luxury, comfort, and good health than ever before: our ancestors could scarcely imagine that such a large swath of the population in the Western world, the emergent middle class, could enjoy such opulence. But the middle class in America emerged only quite recently (after the Second World War) and lasted scarcely fifty years before being undermined by deregulation and greed. More and more it appears to be a transient historical anomaly: at this point the luxury of leisure granted by the Global Economy mostly benefits a few plutocrats, just as it did prior to the twentieth century (Hedges, 2009; Wolin, 2008).

And yet, unlike in previous eras, most of us now depend for our survival on the existence of that Economy, as we lack the knowledge and skills needed to eke out a more 'natural' living. Wendell Berry (2003) has written that the situation evokes an image of a giant airplane full of passengers that has taken off, but has no place to safely land or refuel. The number of human beings may well be more than the earth system can support without industrial augmentation fueled by non-renewable energy reserves that animate high-tech machines whose operation depends on highly refined, non-renewable minerals extracted from the earth. In short, humanity has developed, and thus become dependent on, a system whose persistence depends in large part (if not entirely) on the petroleum-fueled metabolism of global economics, which can only be sustained by mechanical repair systems uniquely enabled by that same metabolic system.

What does such repair entail? For one thing it requires an organized industrial infrastructure that relies on the development of increasingly sophisticated technology. The human population explosion has been accompanied by an explosion of information, and to meet this need, an exponential increase in information processing capacity. At the current moment in history nothing in the civilized world gets fixed without computer assistance. Technology development continues apace, leading futurists to wax poetic about a 'singularity' event in the near future, when our cybernetic technology takes on a life of its own by merging with, and thus transcending, human intelligence. Depending on your perspective, such cyborgation of humanity is either the ultimate fantasy or the ultimate nightmare of science fiction. However, as we witness the beginnings of global economic and social collapse on an unprecedented scale we may want to consider that this particular scenario may be more fiction than science.

In the Global Economy, repair requires growth, as it is largely accomplished through replacement coupled to 'externalization' of waste. When something wears out we simply throw it away and purchase a replacement. This purchase then has the dual function of effecting a repair and stimulating economic ('metabolic') activity.

To be sure some forms of repair that are less wasteful than others, involving only replacement of broken parts with newly manufactured ones (e.g., as in car repairs). But for the most part we participate in a disposable economy, which depends for its growth on us buying new things, leaving the old worn out things (and the toxins they eventually release) to litter the countryside and accumulate in dumps, landfills, the atmosphere, and enormous ever-growing patches of garbage in the world's oceans. Growth of the consumer economy—and hence accumulation of wealth (capital)—depends on this wasteful 'repair through replacement' way of life (Magdoff & Foster, 2011).

Over the past four or five decades progress has been made in recycling of materials, reducing the amount of waste and this increasing the efficiency of the system. But production and consumption of disposable materials remains far more profitable, because in the Global Economy the cost of environmental waste accumulation and human suffering (much less animal suffering) is simply ignored and passed on to those who lack the power to refuse—which includes future generations, who will be deprived of many of the resources we take for granted. And so this burden of 'externalized' debt grows unabated, increasing stress on the system. As we have seen, systems eventually reconfigure in order to relieve mounting stress. In human societies, such reconfigurations are typically violent and cause immense, widespread mortality and suffering.

On top of and intertwined with the economic development is the technological development that affords continued growth, and hence repair through replacement. As noted above technology development continues to accelerate, with no signs of stopping (Hayles, 1999; Arthur, 2009). It is still in a growth phase, a testament to the deep well of human intellect and creativity, and the power of mechanistic models. This inspires the commonly-held, faith-based belief that technology will solve the very problems that its economic deployment has created, such as resource depletion. And perhaps it will. But history suggests (and the Second Law demands) that all technological solutions come with their own problems, and given the current state of socio-economic affairs, one has to wonder how much reserve capacity we will have to deal with those problems.

An important function of technology is to facilitate communication. In addition to contributing to the development of metabolism and repair, this has had two significant effects, the first being the transformation of politics.

The second (which feeds back into the first) is that much of our leisure is now dominated by the medium of television, and more recently, the internet. These are not trivial developments: they have played a central role in the evolution of contemporary civilization. The technology through which we obtain, transfer and process information interacts with a human brain that remains developmentally plastic throughout our formative years, if not lives, and thus represents a significant new aspect of psycho-social-economic development (not to mention source of *distraction*). Hence we are being changed over the course of individual lifetimes by the very technology we created.

So, to recap, the story of the Global Economy is the story of America's development. It began with a growth phase, involving conquest of undeveloped land 'in the name of God', which granted enormous economic privilege and thus advantage to those who participated in the conquest (Diamond, 1997; Singer, 2009). Given that the initial immature stage of any developmental trajectory affords lower-level constituents of a system degrees of freedom not available at later stages, it is not a coincidence that America, after winning its independence from Great Britain, became known as the 'land of the free'. With abundant land, anyone with an able body and sufficient know-how could live in relative freedom. Of course, in the infancy of the United States the land holdings of some men were, either by birthright or through political favor, more abundant than those of others, and so even at that time Thomas Jefferson's ideal of equal opportunity for all remained largely unrealized. Nevertheless, for a few decades there was ample opportunity for 'yeoman' farmers to eke out a relatively independent (i.e., 'free'), if not leisurely, existence on land previously cleared of hostile natives.

But the wealth (and hence power) afforded by land is limited by the power of the system used to extract that wealth. And so it was inevitable that Industrialism would win the day, and that the United States would thus increasingly realize the European model that Jefferson so despised: an urbanized nation, whose citizens are removed from the land, dependent on industrial manufacturing and beholden to banks and other organizations that control the flow of currency. Through industrial development, America, following the rest of the Western World, became organized around urban manufacturing centers that offered gainful employment to specialized laborers. Farmers became more and more dependent on these centers, which developed technologies that increased yields, allowing larger tracts to be farmed by fewer people, who became increasingly dependent on the economic system. So

America's growth, which proceeded from East to West, led directly to its self-organization along the same westward trajectory, a development that realized the industrial model engendered by the mechanistic worldview of the European Enlightenment.

Like any development, that of the United States was anything but smooth: it involved many growth pains, moving in fits and starts through multiple plateaus defined by metastable systemic constraints and internal conflicts. By the end of the nineteenth century wealth had become highly concentrated in the hands of a few 'robber barons' and 'captains of industry' who owned and ran the banking system with little regulatory oversight. For whatever reason (and as with any complex system, there were many), the system became unstable and crashed in 1929, precipitating the Great Depression. This did not end until World War II, which brought about a dramatic revival of the American Economy, owing to the war-engendered self-organization of the populace (a fervently united reaction to the sinister 'other' of Germany and Japan) and subsequent growth of new (and renewed) markets afforded by the Allied victory and America's newfound global presence. This resurging industrial growth resulted in the burgeoning of the 'middle class', a reinvigorated Global Economy, and the baby boom.

The last stage in America's development began in the 1960s, with a new awakening of consciousness that engendered a strong reactionary backlash that continues to this day. Without getting into the political effects of the increased awareness and humanistic ideals embraced by the countercultural movement of that decade, and of the traditionalist reaction thereto that facilitated the election of Ronald Reagan in 1980 and continues as a major source of inspiration for the American Republican Party, we can trace at least three converging trends leading to the economic collapse that is currently underway:

First, there is the depletion of resources, as epitomized by peak oil (Heinberg, 2004), but including as well the depletion of aquifers, minerals, fertile topsoil, and ecological capacity (species abundance and diversity). For the things upon which human life most depends, supply is dwindling as demand increases unabated. Of course, this sort of thing has happened repeatedly throughout history, and has precipitated the collapse of many previous civilizations. But what is different now is that capacity is being depleted *globally*, not just on a continental scale. There are no new continents to conquer.

Second, the baby boom in the West is over. Growth of the Global Economy is based on purchasing power, which for most individuals peaks between twenty and fifty years of age. The number of Americans in that age group peaked around 2010, and is now in precipitous decline; as a result, economic growth in the United States is being strongly countered by demographics, just as it was in Japan in the 1990s (Arnold, 2002). Anti-immigration politics will only exacerbate the problem.

Third, deregulation has allowed for many unscrupulous transactions—basically legalized stealing by the rich and powerful—that have produced economic bubbles that afford an *illusion* of a growing economy amidst a reality that is quite the opposite. The housing bubble burst in 2008, and many other economic bubbles remain on the verge of bursting.

In short, the Global Economy—the system that keeps most of us in the Western World alive and well—is stressed to the breaking point (Magdoff & Foster, 2011). Its homeostatic persistence depends on a metabolism supported by rapidly vanishing natural resources and an aging infrastructure that can be repaired through technological advances only as long as that metabolism remains viable. It is patently obvious that this system is not sustainable. And yet, with the clock ticking we see widespread retreat into cognitive denial and the blissful ignorance of fundamentalism, a form of societal insanity that only exacerbates the 'wicked' problem with which we are now confronted. This makes for a dire predicament. Just how dire is the subject to which we now turn.

# Chapter 6

# RUNNING ON EMPTY

Let us be clear: although we, as a species, are visiting great harm on the biosphere, we are *not* destroying the planet. We are not even destroying life. The late George Carlin, in his inimitably hilarious and penetratingly honest way, made fun of the arrogance that it takes to believe that we have that much power. While we are indeed contributing significantly to the destruction of species, extinction is nothing new—as noted in Chapter 1, it is the evolutionary rule rather than the exception. As a product of nature we are just one more natural cause of extinction, in a long succession of natural causes. But life itself will endure, as it always does, in one way or another. It may not be life as we know it. Many wonderful, magnificent creatures, some never known and others beloved by humans, will soon be gone forever. But we are not destroying the planet.

What we are doing, by way of our mechanized Global Economy, is destroying ourselves.

Even so, chances are that some of us will survive the impending collapse. The question then is: who will survive, and what kind of world will they inherit? That very much depends on the particulars of how things play out, which cannot be predicted. But if past behavior and current trends are any indication, it is highly doubtful that we as a species will come together in harmony to work things out for the betterment of all. Sadly, what seems all too likely—what we are in fact already seeing (yet again)—is that we will turn on each other in violent conflict. Under such circumstances the 'fittest' who survive tend to be those 'red in tooth and claw'. As resources become scarce and inequalities in their distribution grow, culture wars escalate into actual wars.

Nuclear weapons are still very much with us, and their unleashing (be it intentional or inadvertent) could easily wipe out most life on the planet, at least for a while, much as happened in the prehistoric past with asteroid impacts that caused mass extinctions. Overnight the world would become all but uninhabitable. Cormac McCarthy's *The Road* (2006) is a chilling tale of just such a world, in which those who survive are left to cope however they can with virtually no food or water. What makes that novel so frightening (and poignant) is its realistic appraisal of human nature.

Fortunately, we and the world are complex systems, and have capacity for creative change. Outcomes that are highly probable at the moment may never be realized, as what is likely now may be unlikely tomorrow. We as a species clearly have the power to change the world, for better or for worse. But to improve our prospects for the future we need to have a clear, *realistic* picture of what the world is like at present, and knowledge of what happened in the past to bring us to the present moment. This is why we need science, even with all its limitations and practical failures. And science has already given us a pretty clear picture of our predicament.

What we need however is *not* a particular kind of science, much less a particular kind of technology. What we need is the cognitive process that allows us to distinguish facts from fantasy, reality from ideology. This is not as easy as it sounds—most scientists even have a hard time with this. Indeed, most people are more than happy to reject any fact that flies in the face of their favorite theory or fantasy. This follows from the fact that very few people, even scientists who should know better, recognize that the scientific enterprise is itself a living system of models, a social-psychological instantiation of the Modeling Relation. As such it is a subjectively creative endeavor. While science is grounded in facts explained by rigorously conceived, *testable* ideas (theories) about nature, it is not simply an infallible algorithm for obtaining 'objectively' realistic information.

Moreover, as we noted in Chapter 3 it is a mistake to think that some kinds of science are inherently better, or 'harder' (more accurate or precise), than others. For example, psychology and sociology are often disparaged as being 'soft' sciences, and even biology is considered by many to be 'softer' than physics. An opinion held by many non-scientists—and even some scientists—who live in fervent denial of realities such as evolution and climate change is that some kinds of science (evolutionary biology, climatology) are not really science. Similarly, many mistakenly believe that we will ultimately find purely physical (i.e., mechanical) explanations for everything in nature—

and that all problems have technological solutions. Such beliefs are the ultimate reductionist fantasy, based in the mythology of mechanisms that we discussed in Chapter 2. This attitude is not in any way a solution to our problems—it is, in fact, a big part of the cause.

No, in saying we need science what we mean is that we need—*among many other things*— the scientific approach of making sense of sensory information. To do this realistically we need to understand, and make judiciously creative use of, the Modeling Relation and the hypothesis testing that it entails. In other words, *we need a better model for life itself*, which starts with rejection, or at least an explicitly articulated circumspection, of the machine metaphor as an explanation for life.

We will return to this important point in the next chapter. But first we will take stock, and review what science has already made clear about the present state of affairs.

There are of course myriad existential threats confronting civilization and humanity, many of which we have already touched on, but we can focus on one that is particularly troubling because it is actively denied by a politically significant percentage of the American population (who are in fact largely to blame): anthropogenic climate change. Our intent here is not to review the irrefutable evidence that climate change is happening (one need only travel to polar or alpine regions to witness it first hand), or the all too compelling reasons to believe that human beings are playing a significant causal role in that reality, as the case has already been eloquently and quite convincingly made by others (e.g., Stager, 2011; Mann, 2012). Moreover, from the scientific perspective of thermodynamics, anthropogenic climate change is a no-brainer: it is to be expected that the atmospheric dissipation, over the course of mere decades, of energy that took millions of years to concentrate and store in fossil fuels would severely stress and thus change the climate. Scientists who deny this reality either need a refresher course in basic physics, or have sold out and are being dishonest.

No, the problem is not that more evidence or better models are needed; it is that more evidence and better modeling cannot overcome the widespread denial of reality that prevents concerted efforts at remedial action.

But even if we are able to overcome the widespread denial, it is highly unlikely that we will be able to prevent climate change from happening. For one thing, a drastic powering down of the Global Economy would be necessary (Heinberg, 2004), and given the dependencies discussed above we know of no way of doing this without a global movement toward 'shared sacrifice'. Such a movement appears to have a snowball's chance in hell of gaining political traction. Thus, it seems likely that, whether or not a majority of people begin to accept their own complicity, the Global Economy will continue to chug along on fossil fuels, causing the earth's climate to change drastically and irreversibly over the next few decades.

But this will only add further stress to the Global Economy. Among other things it will produce severe draughts and flooding that result in food and water shortages, at first in poor, densely populated regions that can least afford them, but eventually in much of the developed world. It will also damage urban infrastructure, particularly in coastal regions (imagine many more Katrinas, Deepwater Horizons, and Fukushimas), but also elsewhere due to increasingly destructive weather systems. This does not bode well for an economy that is barely sputtering along, already stressed to the breaking point.

How many more toxin-spewing environmental catastrophes can the system withstand? No one knows. As we were writing this a rare magnitude 5.8 earthquake occurred near the Lake Anna nuclear power plant in Virginia, which was followed within days by hurricane Irene. Fortunately, a Fukushima-type disaster was averted—but not by much considering that the Lake Anna plant is built to withstand only a magnitude 6.2 earthquake, and like Fukushima lacks adequate containment or backup power to keep water pumping (necessary for cooling spent fuel rods stored at the plant) in the event of power failure. We were fortunate indeed—a large region around the Fukushima plant is now uninhabitable owing to radiation levels that are hundreds if not thousands of times greater than those released by the atomic bomb at Hiroshima. And the full extent of damage from that catastrophe is still not known.

Climate change, and the destruction it brings, is happening, just as it has happened (owing to various non-human causes) many times in the past. But this time we are (part of) the cause, and more aware of what is happening than our animal ancestors were. So the question is not how will we prevent climate change—that ship has sailed. The question is how (and to what extent) can we control it—and thus prevent overwhelming catastrophe caused by runaway positive feedback—while at the same time adapting to its devastating ecological and economic effects?

As serious a problem as climate change is, it is but one of many environmental hazards we have created for ourselves. We seem to have forgotten, or simply choose to ignore, that in life what goes around comes around. Take plastic for example. This incredibly versatile material is used in just about everything that humans manufacture, and has been for the past six or seven decades. Unfortunately, while it does break down, it doesn't go away. And its breakdown products are toxic, causing numerous health hazards that are now known (e.g., endocrine disruptors such as bisphenol A), and undoubtedly many more that are not known. Plastics and their byproducts are accumulating in the environment—including throughout the world's oceans, where they are being consumed by microscopic animals at the base of the food chain—and are now found in the bodies of most human beings.

Norman Mailer (1963, 1970; Meikle, 1997) railed against plastic in the 1960s, and was ridiculed. He used cancer as a metaphor for the explosive proliferation of plastic and its infiltration into all corners of the world. He spoke of its deadening effect on our senses, of how it forms an artificial barrier between us and nature that makes us *senseless*, and in unconscious reaction, more violent. Now we are learning that environmental plastic likely contributes to the development of actual cancers, and much of what Mailer wrote can be viewed as prophetic. But his intuition was deemed 'non-scientific' because it did not fit the Reductionist paradigm.

And plastics are only one of *many* pollutants that human activity is releasing into and distributing throughout the environment. We have poisoned, and are poisoning, the water and the air. The problem is that living systems did not evolve with the refined materials that we create with our industry (and now depend on), and thus are not equipped to deal with them physiologically or ecologically. While the long term effects of this cannot be known—life may well adapt—over the short term it undoubtedly increases stress, not only physiologically and ecologically, but also socio-economically, by increasing morbidity (for example, increased rates of cancer and congenital defects) and hence healthcare costs.

The problem of healthcare has received a good deal of press in recent years; unfortunately, most of the discussion completely misses the mark due to faulty framing. The billions of dollars spent by and for the biomedical industry (a complex socioeconomic system of pharmaceutical companies, insurance companies, medical practitioners, government, and academia) don't even begin to address the root causes of most contemporary human health problems. Most proposed 'solutions' only treat symptoms of the real problems, because by and large they fail to acknowledge the fact that the environment (which includes all that we eat, most of which is refined to toxicity and sorely lacking in nutritional value) is a fundamental determinant of human health. Ironically, at this stage of its development the biomedical industry, by way of the significant contributions it makes to pollution, climate change, resource depletion, wealth disparity, etc. may well be doing more harm than good for human health.

After its decades-long preoccupation with molecular genetics, biomedical science is only now beginning to understand the profound influence of the environment on human development (Agin, 2010). This is an important perspective that is all but ignored in most discussions of the impending healthcare crisis. For present purposes suffice it to say that what you don't know *will* hurt you. And yet, what we are now witnessing is a strong movement in the wrong direction, with the ongoing political backlash against all environmental regulations and protections. The abuse of science in the name of economics and politics is endemic to the human socio-economic system, a hallmark of the Global Economy (Agin, 2006; Mann, 2012).

Fortunately, complex systems are resilient to stress. Unfortunately, we don't know how far this one (the Global Economy) will stretch before breaking.

But environmental stress is just one side of the coin. On the other side we have the developmental problem of 'senescence'. Stanley Salthe (1993) argues that developing systems eventually enter senescence as a result of "information overload", which makes sense when you consider that development *creates* information (asymmetry, or constraint on flow), through the interplay of growth, self-organization and the creative realization of a system's internal models in the world at large. Eventually however the amount of information created by a developing system (which includes its waste) exceeds the processing capacity of the system's internal models, which, having been realized under growth-favoring conditions experienced during the immature stages of the system's development, are 'geared' toward facilitating growth of the system. In response to information overload a system will often continue organizing around its dominant (most entrenched) models, in attempts to make their realization more efficient by eliminating 'unproductive' overhead (e.g., diverse resources that remain unexploited) (Ulanowicz, 1997). This only accelerates the system's demise, because such actions remove whatever capacity for creative change the system still has.

As discussed in the last chapter, one of the common features of collapsing societies (Diamond, 2004)—and indeed, of dying systems in general—is continued reliance on growth-phase models (genetically encoded 'programs' in the case of organisms) after they no longer fit the system's actions to its circumstances, and have in fact become self-defeating. We can see many examples of this in contemporary civilization.

The most conspicuous example is the Global Economy itself, which relies on growth to generate capital that fuels further growth. The problem here is not 'growth' *per se*—something is always growing, somewhere—but rather the growth of certain human efforts within the context of the contemporary world: of corporate industry, which make products for human 'consumption', and of the human population, which is needed to consume the products created by the corporations. The two feed each other, and this works as long as there are plenty of 'cheap' untapped resources available to exploit: fossil fuels, land, water, and ever increasing numbers of people. Under such circumstances, which existed for much of our history (and continue to exist in some corners of the world), the model works to promote growth, to the benefit of many people in the urban 'middle class' (although many others—those who are powerless and typically 'out of sight' and hence 'out of mind'—are, as noted above, exploited in the process). Circumstances have changed however, as they always do; for various reasons that have already been discussed, continued growth of large corporations and of the human population is becoming increasingly self-defeating. But the Global Economy *has* no other model. And so it (via us by way of our unconscious mental models) continues working to stimulate its own growth.

As a result corporations continue to enrich themselves (meaning, at this point, a relatively small number of individuals get filthy rich), and more and more people suffer the consequences while continuing to feed the corporations, upon which they have become utterly dependent (Reich, 2007; Soros, 2006, 2008; Hedges, 2009; Wolin, 2008). It is a vicious cycle, one that is extremely difficult to break from within, from the bottom up. Doing so requires awareness and concerted effort among large numbers of like-minded people. Unfortunately, there appears to be very little awareness, even among educated people, and a large segment of the population lives in outright denial. The economic model has become *cognitively entrenched*. And so the prospects for breaking the cycle before it destroys itself (i.e., before it destroys us) are not at all good.

This cognitive entrenchment manifests in many ways. Take for example traditional or 'family' values—a catchphrase that in the United States has come to mean *unquestioning deference* to religious (particularly fundamentalist, i.e., literalist) authority, and translates economically to the Protestant work ethic. The latter engenders the belief that hard work is all that is needed to climb the economic ladders to wealth. Conversely, if you are not wealthy it must be because you simply are not working hard enough. While this line of reasoning may sound ridiculously naïve (for example, it mistakenly assumes that ladders of entrepreneurial opportunity are equally available to any and all), it is in fact what many people fervently believe. And they can almost be forgiven for that, because to a large extent it held true in the 'land of opportunity', at least until fairly recently. Throughout much of the history of the United States, when the country was still in its growth phase, opportunity abounded, and education was the major impediment to advancement. Of course, opportunity for education was not equal for much of that period (being restricted mainly to economically privileged white boys), but that began to change as a result of progressive political efforts to equalize opportunity in the twentieth century.

But times have again changed. Hard work and education are no longer a sure ticket to fortune, because opportunities for 'advancement' have all but dried up. We no longer live in a world favoring growth of economies based on industrial production and mass consumption of manufactured things. In the lingo of cell biologists, conditions have become *growth limiting*. So 'traditional family values' that worked reasonably well to promote growth under conditions that favored growth—values such as the protestant work ethic, unrestricted reproduction (and its corollary, lack of freedom for women), and the ideal of a nuclear family with only one parent working outside the home—are no longer automatically rewarded. What is rewarded is as always some combination of talent, increasing specialization in an established field (which, to be sure, requires hard work and education), a charming personality and knack for self-promotion, good luck, and unscrupulous gaming of the system. So those who 'get ahead' tend on the one hand to be highly talented, hard-working (indeed *driven*, often to the detriment of family) and to some extent lucky individuals, and on the other, highly intelligent crooks, bullies, and sociopaths.

And, as always, the latter tend to accrue a good deal of political power, which allows them to stack the deck in their own favor. One of the ways that this is done is through effective manipulation of the power of myth (i.e., by taking advantage of cognitive entrenchment) which allows them to use an age-old and ever-effective strategy: divide and conquer. The way that this has been done in the United States has been to use 'traditional values' as a polarizing wedge to pit the disempowered against each other. This keeps the powerful in charge, while the powerless blame each other for their problems, including their lack of power. Those who have *always* been disenfranchised by 'traditional family values'—for example women and various minorities—tend to blame those values as being *the* cause of their powerlessness (when in fact they are only one aspect of a complex problem), only to get blamed in return by those who *used* to benefit from those values, but for one reason or another no longer do. The rich and powerful, on the other hand, benefit greatly from this culture war, which they stoke through fear mongering and effective deployment of mass media. And those among them who are actually (cynically) aware of what they are doing (many are not, being just as deluded as most everyone else) laugh all the way to the bank.

As far as we know the political system that affords the most power to the most people, and offers the best opportunity for social improvement, is democracy. But the foregoing discussion raises an important question: how effectively can democracy actually work to improve things when the behavior of the majority—on all sides of various issues—is entrained by deeply entrenched, maladaptive cognitive models reinforced daily by a barrage of electronically disseminated propaganda? This brings us back to the problem of framing discussed by Lakoff and touched on in Chapter 2. Without revisiting that particular discussion here, suffice it to say that a majority of people in any society can easily be misled by savvy politicians backed by the corporate owners of mass media—a lesson we should have learned from Nazi Germany, but apparently did not.

In short, when a majority of the population is entrained by cognitive models that are no longer congruent with reality (i.e., when those models no longer anticipate what is *actually* needed for the long-term well-being of the population), one has to wonder whether 'rule by majority' can be effective at changing a system, born of those very models, upon which most people have become dependent for their very survival—even if that system is rapidly bringing about its own demise.

Which brings us full circle to the question of how we got in to such a fix: why are our cognitively entrenched models so maladaptive? That they can reasonably be referred to as models indicates that they are (or were) congruent with some aspect of reality. So why are they leading us down the primrose path to self-annihilation?

The answer emerges from all of the above, and can now be summarized, as follows:

Life, which is in essence a *relational* phenomenon, manifests the Modeling Relation. Systems at all levels (organisms, ecosystems, social economies) are *subjective selves* that, in striving to persist, relate to and interact with each other within a larger system (the biosphere). Those interactions are guided by models, created whenever a system relates to and anticipates events in the 'external' world by encoding specific aspects of their causation, interpreting those encodings by way of formal systems of *entailment* 'internal' to the system, then decoding the entailed outcomes into action that fulfills the needs of the system. However, compared to the actual world, which is complex, all models are simple. They nevertheless serve for a time to fit a system to its circumstances, helping it prepare itself and do what it needs to do in order to persist. But circumstances change as a system develops. Models that work well early in development, when a system is growing and circumstances are conducive to that growth, often become maladaptive later in development, when conditions become growth-limiting. Under the latter circumstances, a system's continued efforts to grow under the influence of entrenched models lead to its senescence, a condition of extreme fragility poised for collapse.

There is abundant evidence for this being the case in individual organisms, in which aging is accelerated by growth-promoting genes and processes (e.g., consuming food *ad libitum*), and delayed when those same processes are held in check (e.g., by diet). From a thermodynamic perspective this is not at all surprising: growth requires a relatively high rate of energy consumption, which by the Second Law entails a high rate of waste (entropy) production. If growth-promoting processes continue to hold sway in a system that lacks the capacity to grow, then that waste—much of which is toxic—is bound to overcome the system.

But this is not widely acknowledged because science has worn the blinders of reductionism. This focused outlook worked well for a time—the mythology of mechanisms afforded an effective model for the early development of Western civilization, promoting its rapid growth following the European Renaissance and Enlightenment and during its subsequent conquest of America and other undeveloped continents. But that scientifically informed mythology occludes consequential context, leaving a cognitive void that attracts superstition.

And yet, context is everything, and key to complex causality. The fallacy of reductionism is that context can be fully 'comprehended'. It cannot, because it is not possible to comprehensively 'view' a natural system from 'outside', or even to isolate it in order to perform a fully 'controlled' experiment. Objectivity is a myth, and there is no largest model that fully encompasses context. To think otherwise is to mistake fantasy for reality. As a farmer once remarked, "that's just plain foolishness."

Many so-called complexity scientists acknowledge the limits of reductionism, but nevertheless fail to recognize that science and technology alone are inadequate to that task of solving the problem with which we are now confronted—and indeed, are part of the vicious cycle that is causing the problem. Such scientists are disinclined to entertain views such as those so eloquently expressed by Wendell Berry in *Life is a Miracle: An Essay Against Modern Superstition* (Berry, 2000). All too often scientists use their credentials as an authoritarian club to beat down such criticism, much as was done in response to Rachel Carson's *Silent Spring* (1962), or to Norman Mailer's campaign against plastic. And so the problem cyclically worsens.

It does so because life is not a mechanism that can be adequately explained in terms of external actions or 'forces' that efficiently *cause* some reaction—or even in terms of the mathematically or computationally tractable dynamics of systems. Rather, it is a self-entailing aspect of the world (or universe), which means it causes itself, from within. And as it does so, it develops.

Development establishes context that defines what can change, and the range of available choices. It is teleological (manifesting final cause) because it is driven to anticipate by the needs of a system. In disallowing consideration of consequences (final cause) as a legitimate concern of scientific inquiry into causation, and claiming that the inner workings of nature can be adequately explained in terms of mechanisms, Bacon, Descartes, *et al.* blinded the scientific enterprise to the consequential nature of developmental context, and in so doing set us on our current path of self-destruction. Mechanisms, which are but one aspect of the natural world, became the be-all and end-all for science, which in turn became an unwitting means to an end: economic growth through efficient exploitation.

While the mythology of mechanisms was adaptive during growth phase of the Global Economy, it is no longer so: the system has developed past maturity and economic growth based on mechanistic objectification is no longer adaptive, and is instead self-defeating. It is analogous to how processes that promote growth of an organism in its immaturity become unhealthy (e.g., oncogenic) after the organism has matured.

Anyone who is close to nature can see that there is no such thing as unlimited growth. In nature, systems either grow to a limit and die, or find a way to achieve a homeostatic (ecological and physiological) *balance* of continuous growth offset with continuous death: the ultimate recycling program. It doesn't take a formal education to be aware of this organic reality, and perhaps those without such an education can see it more easily than those who learn about the world from books written by 'experts' trained in the mechanistic tradition. Berry is right: small (non-industrial) farmers know a good deal more about real life than most scientists.

In putting our remarkably acute scientific models to work in transforming our world we disconnected ourselves from the nature of life itself, and thus lost touch with reality. A word commonly used to describe such a condition is *insanity*. This is an apt word, because 'sane' comes from the Latin *sanus*, which refers to health. To be sane is to be of sound mental health, whereas to be insane is to be of ill mental health. By this definition, the self-organized system that manifests our collective mentality—the Global Economy upon which we have come to depend for our survival—is insane.

What, if anything, can we do to get better? We close this essay with some suggestions.

# Chapter 7

# WHAT CAN BE DONE?

So far we have discussed how human cognition, empowered by formal logic and scientific knowledge, transfomed the world by fostering development of an industrialized global civilization, which has now developed beyond the limits of sustainability to the brink of catastrophic collapse. And yet, most of the developed world goes about its business as usual, not just continuing, but actually accelerating the very activities that brought it to the brink. We have asserted that this can be reasonably described as, if not attributed to, a form of widespread insanity, manifesting an unbalanced mentality and cognitive models that are incongruent with reality.

We have thus offered a diagnosis. Can we prescribe a cure?

It would be presumptuous to say that we can. The best we can do is to try to provide our own perspective, and offer our opinion on what we consider to be realistic options. We suggest that a necessary step is acknowledging the insidious influence of Reductionism, the mythology of mechanisms. We need to ask: how does it unconsciously guide our actions? And what, if anything, can we do about it?

That mythology, realized in our mastery of machines, allowed us to develop technology of unprecedented scope and power that has become an entity unto itself (Kelly, 2010). In so doing we lost all perspective of context and limits. We *do* because we *can*, with very little awareness of, much less regard for, the actual (as opposed to the ideally intended or fantasized) consequences. And in that we have become unconscious slaves to what we have created.

Change on an individual basis is hard enough, as anyone who has struggled through therapy knows. So it would seem well nigh impossible for us, collectively as a species, to step outside ourselves and critique what we see when we look at what we have become. And that is the problem: we must bring it all into consciousness if there is to be any hope of using our unique gift of rationality to make healthy choices—or of recognizing that a significant part of our problem is a habitual over-commitment *to* rationality, at the expense of intuition and empathy.

In Chapter 4 we discussed how continued development of any specific thing—*rationality* for example—ultimately leads to the demise of that thing. In exploiting specific opportunities, development increases the level of commitment to (and hence dependency on) specific ways of being, and while that can be good, it can also be bad, depending on the context. Moreover, it can change from being good to being bad when the context changes, as it generally does. Insofar as the increased level of commitment needed to take maximum advantage of a specific opportunity bars access to other opportunities, development can be thought of as a trap. We have tried to suggest that we can better appreciate our current predicament by reflecting on what has been lost in the development of Western civilization. This of course is the flipside of the usual focus on what has been gained. But balance requires awareness of both sides.

As we have tried to make clear, sensory information is meaningless without a memorized framework in place to interpret that information. Cognitive frames for interpretation, discussed at length by George Lakoff, are assimilated into our unconscious. Such assimilation is itself a developmental process, occurring at many levels, from that of individual psychology on up to the socio-cultural and the 'collective unconscious' described by Jung. At the individual level this development is not unlike what happens with master musicians who, after repeated practice, become able express music unconsciously and hence more freely. Such development can also be seen in skilled athletes. What is less obvious is that all our education and learning similarly becomes assimilated into unconscious, habitual ways of relating to the world. While that has its benefits, one would expect it to be problematic if the education and learning are based on and reinforce misconceptions that disconnect us from the real world.

Since our interpretation of information of any kind depends entirely on the socio-psychological development of cognitive frames, our minds are totally subjective. It takes a conscious awareness of that subjectivity to recognize that what we call 'objectivity' is an unrealizable (and hence unrealistic) ideal. That recognition is in turn a pre-requisite for seeing that we are naturally prone to (and might well be, even in our most 'rational' moments) unconsciously and habitually *mis*interpreting input from 'out there'.

This gets us back to Francis Bacon's noble effort to free us of the subjective 'Idols' that bias our perceptions. His solution to the problem was the Scientific Method, which uses controlled experimentation to make everything as objective as possible (but note the implicit subjectivity in the fact that the controls must be chosen!). Toward that end, final causes, deriving as they do from consequences that are mentally (and hence *privately*) anticipated by a subject, are disallowed as objective (i.e., 'scientific') explanations of the natural world (Rosen, 2000). To a point this is well and good; however, as we have seen, it is an unrealistic approach to the living world. And it creates serious problems when it is combined with Cartesian dualism and Newtonian determinism and internalized as an unconscious mental frame that mistakenly yet habitually conflates the ideal of objectivity with mechanistic causation, giving rise to the presumption that the world is nothing more than an objective mechanism that can be controlled at will, given sufficient technological knowledge. When that happens we fail to notice that the real world is complex, which means that there is a subjective aspect to everything—at least to everything that matters. A casualty of that failure is empathy, the very foundation of morality.

So it behooves us to come to grips with the fact that objectivity is an ideological myth that can never be fully realized in the actual world. To pretend that it can is to take a step away from reality. This does not mean that objectivity is not something we should strive for (as much as possible in appropriate contexts), only that it is something we need to put into realistic perspective. Toward that end it is useful to ask how 'objective' ways of relating to the world develop ontogenetically, and how they have developed in human evolution. That question gets beyond the scope of this essay and we will not try to answer it here. Suffice it to say that objectivity is not something we are born with, and not something that our animal ancestors had. As far as we know only human beings are capable of viewing the world 'objectively'. According to Iain McGilchrist (2009) this faculty co-develops with the technologically manipulative mode of attention that is the specialty of the left cerebral hemisphere—a mode of attention that works by virtue of the language-enabled cognitive distance that it *interposes* between the self (subject) and the other (object). But for that to happen a *choice* must be made where to draw the line between self and other, a choice that can only be made by way of interpretation informed by pre-formed (e.g., linguistically entrained) cognitive frames.

So in regard to 'objectivity', the key, yet seldom asked, questions are: where do you choose to draw the line, and why? We suggest that asking and attempting to answer those questions reveals the subjective nature of any such line, and is a necessary step back toward reality.

It is often noted that the existential crisis with which we are faced manifests addiction. In individuals, addictive behaviors often co-develop with psychological defense mechanisms that serve to maintain the addiction by keeping it buried in unconsciousness, giving it a 'life of its own'. These include denial, the rejection of facts that are disturbing to the addict; rationalization, the post-facto creation of plausible but false explanations for one's actions; and projection, attributing one's own issues to someone or something else. So instead of working to free themselves of addiction, addicts will typically deny that they have a problem, or that they are causing problems for others, and blame others for whatever hardship they are experiencing as a result of their addiction. Often an addict will not admit that the addiction is a problem until it causes them to lose everything.

How applicable is the term 'addiction' to what we are discussing here? Are we, in our collective psychology, addicted to how we have come to live? Has the Baconian-Cartesian-Newtonian mythology of mechanisms engendered development of addictive cognitive frames—comfortably entrenched worldviews and ways of thinking—that keep us from making viable choices? Are we terminally addicted to the 'creature comforts' afforded by our unsustainable, earth-despoiling, machine-augmented way of life? We submit that this is where we start.

One of the things recognized by programs that treat substance abuse in individuals is that there is a social component to the problem—that is, the addiction is *enabled* by developed social structures. Substance abuse treatment programs also recognize the need for the addict to admit their problem. In our addiction as a civilization we are nowhere near that point on any scale that has a significant impact, even though some of us are beginning to recognize that we have a problem.

To many of us, it is painfully obvious that our industrialized way of life is violent, unsustainable, and leading inexorably to the collapse of the social, economic, and ecological systems on which most of us depend. However, many (perhaps most) others beg to differ, holding the dissenting opinion that there is nothing inherently wrong with our way of life—as epitomized by George H.W. Bush's famous declaration (over twenty years ago—yet how little has changed!) that "the American way of life is not up for negotiation." This is classic denial, the addict saying "I don't have a problem". It is quite obviously manifest in the widespread denial in the United States of anthropogenic climate change, which serves to reinforce the addict's abdication of responsibility. It is also manifested by the widespread denial of the empirical fact of evolution, which serves the same ultimate purpose by allowing the addict to use the cognitive framework of religious literalism to project blame.

An addiction is simply a well-developed way of being. Addictions don't just happen; they develop over time, like everything else, by way of self-reinforcing loops of causality. So in light of the foregoing discussion concerning the inexorable increase in dependency and determinacy that accrues with development in complex systems, it would appear that at this late stage the probability that a politically effective number of people will willingly choose to change their way of life is slim to nil. Nevertheless, something has to give, so from our perspective collapse of the socioeconomic systems that sustain us is all but inevitable. While this will not come as a surprise to anyone with knowledge of history and awareness of developmental phenomenology, this in no way diminishes the frustration of watching it happen while being largely powerless to prevent it—a frustration felt acutely by anyone who loves children.

The silver lining is that collapse affords evolutionary opportunities, opening up new niches that did not previously exist, removing rigid system-level constraints that suppressed the emergence of alternate models. This too can be seen in myriad examples throughout history. If an asteroid impact had not precipitated the global ecological collapse that caused the extinction of the dinosaurs, then mammals would not have (had the opportunity to) come into ascendancy, and we would not be here.

The dark cloud within that silver lining is that transformation through collapse and rebirth—metamorphosis of the system, as it were—entails, in transition, widespread *death* (Volk & Sagan, 2009).

And this gets to the crux of many matters. Awareness of one's own mortality is the bedrock of human consciousness. The development of Western culture was arguably motivated by the specter of death, a defining fact of life most of us wish were not so. As a result, we have created a world that allows us to pretend that it is not—an artificial reality that engenders the hope that death is a problem that can ultimately be overcome, if not by righteousness then by a technological 'fix', while keeping us preoccupied with multimedia fantasies (most of which are, not coincidentally, sexually charged) that either make the specter easy to ignore, or desensitize us, via a barrage of violent imagery, to its emotional reality. This *repressive* response is epitomized on one hand by the fantasy of an immortal disembodied soul (whatever that might be—it certainly is not obvious given both the complexity and developmentally emergent nature of psychology!), and on the other by Ray Kurzweil's bizarre futuristic vision of a cybernetic 'singularity' that will supposedly allow human beings to evade death altogether.

Let us for the moment consider what the latter scenario might entail, in order to ask first how well it fits with what we have discussed so far, and second whether it is actually something to be hoped for. The 'singularity' is so called because it represents a future 'event horizon' of technological reality, beyond which we lack the capacity to envision. The concept derives from the empirical fact that the rate of human technological evolution—particularly the realm of computational electronics and engineering—has accelerated exponentially, and continues to do so. We have created computers that essentially pass the Turing test (for intelligent behavior) of being able to carry on a conversation indistinguishable from that of a human being. So it does indeed appear conceivable that computer technology will continue developing apace toward unfathomable realms of superhuman machine or 'artificial' intelligence (AI).

But what does that actually mean? The Turing test, named after Alan Turing, the mathematical genius who was one of the founders of modern computer science, turns out to be a can of metaphorical worms.

In considering the power of computation (which is based on mathematical formalism) recall that mathematics opens a limited (i.e., narrow) window on truth. The cognitive tension developed between *this* truth and the failure to acknowledge it is artfully explored in Janna Levin's historical novel *A Madman Dreams of Turing Machines* (2006)—a recounting, by a narrator whose own sanity is open to question, of the lives of Alan Turing and Kurt Gödel, two twentieth century prodigies with divergent views of reality, who both committed suicide.

The key issue here is the nature of life itself, and the centrality of the Modeling Relation. After you absorb the implications of Rosen's insight, the idea that there can be a single form of 'intelligence' that can be developed *ad infinitum* to approach omniscience becomes quite absurd. The rate, capacity, and sophistication of computation have indeed increased exponentially, and this has allowed human beings to construct computers that do a remarkable job of *simulating* human 'intelligence'. But what is it that is being simulated? Any simulation is of necessity based on a model (as opposed to the *actual* world), so what is simulated by AI is the mechanistic (or algorithmic) model of reality bequeathed us by reductionist science. What AI is then is not a simulation of human intelligence (*per se*), but of human intelligence *modeled* as an algorithm embodied in a machine. It should be clear by now that these are not by any stretch the same thing. Unlike simulations, life itself is *not computable*.

As we have discussed, modeling is a developmental process that requires that choices be made, 'trade-offs' that require that some potential outcomes be sacrificed in order to realize others. Development of models can indeed produce mechanistic systems that efficiently cause specific effects, one of which can be the assimilation of information as memory, which (to a point via algorithms) facilitates anticipatory behavior. An invariable 'side-effect' is the dissipation of energy into entropy—both in the strict sense (i.e., unrecoverable heat) and in the more general sense of effects that run counter to the anticipated outcomes, i.e., the telic goal of the designed mechanism. All of these effects entail a loss of capacity or potential, as creative potency is focused, through a series of symmetry-breaking choices that progressively reduce existing degrees of freedom (while creating new ones via growth), toward some particular end. A cost of fully realizing that end is inflexibility, an inability to adapt to all of the vagaries of existence.

So, even if we have the capacity to cross all the energetic and informational hurdles to creating 'cyborganic' machines, the question remains as to whether such technology is sustainable. Developmental phenomenology would suggest that the only way that it could possibly be is through continued growth. Since the capacity for such growth on planet earth is obviously limited, it seems the only way to sustain cyborganic growth would be to discover a way to feed directly on electromagnetic energy, which could afford potential for growth into interstellar space.

As we have seen, whatever technology we came up with to allow that would have to retain the defining attributes of life itself—a truly complex system of metabolism and repair that is closed to efficient cause—in order to be self-sufficient. Machines *alone* lack that capacity. But perhaps human beings can be hybridized with (and thus augmented by) machines to create a new cyborganic form of life.

That may be possible, but would we want to go there?

We will leave that question open, for it has been explored at length in the science fiction literature. Suffice it to say that technological dreams have a recurring tendency to develop into nightmares. We bring it up only to acknowledge the commonly held belief among the scientifically-minded that technology will save humanity. To us this seems as misplaced as faith in the 'external' deities created in the bicameral minds of our ancestors.

The problem with placing one's faith in machine 'intelligence' is that it is a narrow manifestation (realization) of human intelligence, and but one aspect of a well-developed autonomous human (eco)system consisting of interacting causal loops of economy, technology, cognition, etc. As N. Katherine Hayles (1999) has pointed out, human beings are (to a large extent unwitting) parts of that system, and not its master. It is hubristic arrogance to think that we can change that.

The point of this diversion is that awareness of death as a fact of life has created a cognitive need for *faith*, which is readily fulfilled by fantasy—be it religious or technological. One might wonder whether, and to what extent, there might be better ways of meeting this need—that is, of assuaging the fear of loss engendered by our awareness of mortality. We believe that there is. If there is any faith that is not misplaced, it is faith in *life itself*. We shall return to this below.

The specter of death points to the heart of a hard reality that civilized humanity has yet to adequately assimilate into its models: the Second Law of thermodynamics. The Second Law is the reason things wear out, and the reason that life entails death. It explains why things develop but only to a limit;

why overdeveloped systems collapse; and why time reversal and perpetual motion machines are impossibilities. And yet: people still try to invent such machines, still talk about going back in time as if some day it might happen, and still approach death as a technical 'problem' that can be 'solved'.

It is not. We are all going to die, and there is not a damn thing anyone can do about it. The realistic question is not how to avoid or overcome death, but rather, how do we want to face it? And, in the face of death, what persists? And, most pertinently, if not importantly, what do we *want* to persist, and why?

Utopia is not possible. However, human culture has progressed considerably, and while morality is seldom a matter of simple black and white choices, it is easy to identify human (humane) social and behavioral attributes that are desired by most people from a very early age, which form a foundation for a 'civilized' society. In the United States most of these attributes are at least superficially embraced by the 'left' ('liberal' or 'progressive') end of the political spectrum: empathetic compassion for others (as encoded in the Golden Rule), a measure of tolerance, and freedom of expression and choice, constrained by an evolving code of personal responsibility established through reasoned, open-minded discourse. Many on the 'right' ('conservative') end of the political spectrum disparage those values as being idealistic (i.e., unrealistic), or anathematic to traditional (typically religious or nationalistic, and generally literalistic) systems of belief that are held to be beyond question (Lakoff, 2008).

So, whereas those on the 'left' tend to advocate complete freedom of mind combined with social control of economic activity in order to protect against predators and adapt to evolving circumstances, those on the 'right' advocate complete freedom of economic activity combined with social control of mind to prevent its evolution away from traditional models that are considered to be absolute Truth.

As discussed by Lakoff, this difference stems from different conceptions of causality. Conservatives tend to believe that events and phenomena can be attributed to single, 'direct' causes, that once identified, exclude, in binary 'either-or' fashion, the possibility of any other causes; whereas liberals are more cognizant of the fact that causation is complex and multifaceted. Hence the right continues to rely on simplistic (i.e., unrealistic) models—not unlike those of reductionist science. It is not a coincidence then that many on the far left gravitate toward deconstructive postmodernism, which is critical of both science and conservative values.

It should be obvious that the perspective advocated here lies on the 'far left' end of the political spectrum, and we will undoubtedly be criticized for being unrealistic elitists, and demonized as enemies of god and country (not to mention socialists, Marxists, etc.). But to this we can only respond with a question: which perspective do you honestly think offers the best opportunity to create a system that provides for the *common* good and long-term human wellbeing in an ever-changing and now (in human terms) rapidly-deteriorating world?

Once the complex nature of causality is recognized, it becomes obvious that actions based on simple causal models are at the root of the problems we have created for ourselves. While this vindicates many of the values promoted by liberals, the thinking on the left has also fallen into the cognitive trap of seeking simple direct causes. Progressives have misunderstood the role of unconscious thought and framing, and their abiding faith in modernism (the eighteenth century Enlightenment mythology of mechanisms) has led them to believe all that is needed is more data, and rational thought will prevail and all will be well. By and large liberals are just as beholden to reductionist science as conservatives. The irony is that the many hubristic failures of reductionism have only increased the public distrust of science so decried by liberal elites.

So in the end, there is not all that much difference between the left and right, as the full spectrum of American politics is firmly entrenched in a way of thinking that becomes more unrealistic and less conducive to human wellbeing and survival every day.

Which brings us back to the reality of death and the Second Law, and the question: what do we want to persist, and why? The urgency of answering that question increases as we transition into collapse and the end of life as we (have come to) know it.

What we *want* to persist and what we *need* to persist are of course different things. But it is healthy to want what you need, and vice versa—this is part of what it means to be sane. What we need to persist as a species is a biosphere that provides us with food, water, shelter, and some sort of social support system. But there are *many* ways toward that end, any one of which will allow some number of us to survive in some way. Some of these leave more to be desired than others.

So, what do we want to persist?

One way to answer this is to identify things that we do *not* want or need, human activities that are contributing to the current malaise. This part is easy: we do not want or need fear-mongering politics and literalistic religious fundamentalism in the service of mind control, or science and technology in the service of uncontrolled industrial profiteering. Nor do we want or need a system that allows the unscrupulous among us (and there are many) to accumulate an obscene amount of power (wealth) through cunning manipulation of those things.

In deciding what we do want to persist it pays to consider culture. While multiculturalism, relativism and tolerance have been in vogue on the left, the idea that all cultures are morally equivalent is absurd. Most of us (left and right) will agree that some cultures are depraved and better left in the dustbin of history. Cultures that condone slavery or encourage racism are one example; misogynist cultures are another. It is reasonable to view depraved cultures as diseases of humanity that ought to be eradicated.

But how can this be reconciled with the ideal, espoused here, of complete mental freedom? Does not culture extend from cognitive models (encoded in mythologies), meaning that cultural depravity is indicative of mental depravity? In which case, eradicating the former requires eradicating the latter?

And is this not the function of fundamentalism and other forms of mind control?

Yes, but these do not work—they are thoroughly unrealistic. There are much better ways.

Consider again that cultural depravity is indicative of mythological stagnation and cognitive entrenchment—a sign of senescence. Development of a system naturally leads to senescence *unless* the system is periodically disturbed (shaken up) by external interactions. So what is needed to prevent depravity is not mind control, but quite the opposite: a mental way of continually stepping 'outside' one's worldview and considering it as dispassionately as possible in relation to other worldviews. This is why we need an ongoing inter-cultural dialog such as that advocated by Robert Maynard Hutchins (Mikulecky, 2011): to stimulate the dialectic that drives cultural evolution. But (contra Marx) the dialectic can *never* be resolved—there is no ultimate synthesis of thesis and antithesis, because that would be a developmental dead end, and the world never stops changing. And more to the point, as the work of Gödel and Rosen showed, no single formalism or model can ever capture the full truth or essence of a complex system; any complex system can in fact be modeled in an infinite number of ways.

This idea is echoed by Michael Zimmerman (2005), whose Integral Ecology emphasizes that no ecological situation or problem can be understood from a single perspective. Multiple perspectives, from both inside and outside of a given situation, must be taken into account. Each perspective has its own truth-claims that can only be critically evaluated by those trained within that perspective, and which are immune to falsification by the methodologies used in other perspectives. On the other hand, within any given perspective (e.g., science or art) some things are true, and others are not. Since complex systems cannot be adequately described by or reduced to one perspective or another, it behooves us to bring as many different perspectives to bear as possible on complex ecological problems. These would include not only the 'externalist' perspectives of the sciences, but also the 'internalist' perspectives of the humanities (the 'Two Cultures'; Snow, 1998).

Unfortunately, this is not how things are typically done in human societies, wherein the *modus operandi* is instead for a single perspective to become dominant through active repression of others. This happens at the socio-cultural level and extends to the cognitive level within individual minds.

The problem with cognitive repression is that it often has a paradoxical effect of bringing out, in unconscious and unhealthy ways, the very behavior that is being repressed. In fact, in a certain light this can be viewed as its purpose: since socio-cultural systems of morality are not very effective at controlling behavior, they must provide a 'cover' for their impotence. Repression creates a closed cognitive loop that allows, typically through *ex post facto* rationalization, actors within a moral system to believe they have acted morally when in fact they have not.

So, social mind control, by way of repression, is ineffective at preventing depravity. But what other ways might we have at our disposal?

From a developmental perspective, there is no *a priori* reason why a socio-cultural system cannot be constructed that encourages mental freedom (i.e., creativity) while discouraging inhumane and ecologically unhealthy activities.

In an 'ideal' world anarchy would be feasible, and there would be no need for government. And perhaps it is possible that human beings will eventually evolve to the point where law and regulation are no longer necessary because the global consequences of antisocial behavior are widely understood and viewed as unacceptable. In such a culture infractions become the source of their own punishment because of the social alienation one inflicts on oneself. The question of whether this is it possible is age-old, as is the question of whether immoral behavior ('sin') is inherent to being human (e.g., a price paid for the maintenance, in the 'gene pool', of human traits that are needed for survival) or simply a consequence of social immaturity and bad myths: is it nature or nurture (Harris, 2010; Deloria, 2003)?

The problem is that the development of a healthier culture can only be initiated from the bottom up (in this case by individuals). And currently that is quite difficult because of the top-down cultural and industrial constraints that are already in place owing to earlier systemic development, as we have already discussed.

However it might be approached, eradication of cultural depravity cannot be done in a way that is morally depraved. That only replaces one form of depravity with another. But avoiding that requires a fundamental moral standard of personal conduct. In the absence of religion, what might that be? That's easy: the Golden Rule. *Do unto others as you would have them do unto you.*

For most people past a certain stage of developmental maturity, this makes perfect intuitive sense, and everything else follows, both rationally and emotionally.

From the Golden Rule it follows that cultural depravity cannot be eradicated through violence.

Therefore, the only way to improve culture for our descendents is through nonviolence.

This is not easy. Quite the opposite—it is the hardest possible thing to do. Throughout history nonviolent cultural activists have met a violent end at the hands of the powers-that-be. But, that nonviolent activism is perceived as threatening is a testament to its great power, and evidence for the truth of our view that it is the only way to ensure that what we want to persist does persist. Moreover, there are many examples of nonviolent revolutionary change. We saw it recently in the Middle East with the Arab Spring, and are seeing it happen right now in the United States with the Occupy movement. Unfortunately the effectiveness of nonviolent movements is often excluded from our sphere of awareness, owing to the violent context of the system we live in (as we are reminded continuously in the media).

Herein lays the sublime irony: Transformation of a system—its rejuvenation or metamorphosis—often entails death. But the only way to improve culture is through nonviolence. So, morally, the only life that one can be willing to sacrifice for the sake of humanity is one's own.

And there's the rub.

This brings us back to the fear of death being the central problem for human consciousness. The only way to overcome that fear is through faith—hopeful belief that there is life after death.

But this does not require belief in anything supernatural, or that we indulge the fantasy of an immortal disembodied soul. All it requires is deeper insight into life itself. Of course there is 'life after death'. In the long run life improves itself. We may not be there as individuals to experience it, but we can have faith that it does.

We need this faith not only to overcome fear of death, but the repression through objectification that is at the root of so many of our problems. We need faith that, with time, humanity will develop a healthier mythology and culture—a better appreciation for the subjectivity of the world, and thus greater empathy for the *other*.

We need faith in life itself, faith that comes from understanding that all of our social systems—economics, politics, religion, science—are just developmental realizations of cognitive models that are historically entrained by mythology. Such faith comes (at least in part) from awareness that we are each part of a living whole, and that our contribution to that whole is not easily erased, even after we exit the stage. This karmic reality engenders morality and can be a source of great joy.

So, let us again ask: what do we want to persist?

What we want to persist is the 'soul' of humanity.

By 'soul' we mean that which resonates *intuitively* and *emotionally* with the enchantment that Nature holds for those who pay attention. It is that which is revealed in aesthetics—in music, poetry, and art. It is something primal, an instinct that we inherit from the primordial source of our *Being*.

In order for us to reconnect with nature and thus return to health, the human soul needs to be restored and reinvigorated. In other words we need a revival.

Freya Mathews (2003) has argued that the ecological problems that we have created for ourselves cannot be solved scientifically or even rationally— that, in fact, overdevelopment of discursive rationality at the expense of poetic intuition is a significant part of the problem. In her view science cannot provide a fix, because the crisis is metaphysical, not merely technological. Through language, human beings have become entranced with *explaining* the world, and have forgotten how to *relate* to it. In developing our reliance on objective discourse we have repressively reduced our capacity for subjective empathy—not only for other human beings, but all other forms of life. According to Mathews we will only regain this capacity when we (re)learn to encounter the world erotically.

Now this ought to ring a bell for anyone who participated in the explosive elevation of consciousness that occurred in the 1960s. Unfortunately the surge of youthful energy and idealism that helped bring about the countercultural advances of that decade, an erotic awakening that peaked during the summer of love, quickly degenerated into the self-indulgent hedonism that marked the 1970s, giving way to the reactionary cultural backlash of the past three decades. But this doesn't mean that all was lost. Many of us—perhaps more now than ever, thanks to continued attraction of naturally rebellious youth to counter-cultural ideals—remain convinced that, to change the world for the better, "all you need is love".

The point here however is not to romanticize or celebrate the past, which was far from glorious. The point is to remember that there was a time, not too long ago, when Westerners began to (re)awaken to the enchantment that the world offers when it is encountered erotically. It is the same enchantment that most of us feel as children, before we are taught that the world is nothing more than a machine.

And indeed, where there is youth there is hope.

So, a revival starts with education—with teaching your children well, and just as importantly, with *listening* to them. This is not the kind of education most of us are accustomed to. Yes the basics of reading, writing, and arithmetic are necessary, as are history, science, health, physical education, etc. But what is also critically needed, and what is unfortunately missing in so much of modern education, is the nurturing of individual creativity. All too often the first thing to be cut from the public education budget is the arts.

The most important job of education is to maximize the developmental potential of children. This requires that we make them aware of as much of the world as we possibly can while allowing them to find their own way—within the moral boundaries of the Golden Rule—of relating to it. Every child is endowed with unique creative talents, but these are often squashed in the service of mechanical training—the fashioning of gears for the machine of the Global Economy. Education as we know it is destructive, an exercise in putting things into the 'right' boxes instead of seeing how they relate to the whole—a spirit-crushing experience that typically leads to a life of clock-watching employment, material acquisitiveness, unhealthy acting out on repressed impulses, and substance abuse.

Skill training is obviously important, but it is just as (if not more) important that education indulge curiosity and creativity. Music, poetry and art are just as important as math, science, and history.

One of the tragedies of the modern world is that if the industrial food system were to fail (as it eventually will), most people would starve. We are so disconnected from nature that most of us have forgotten how to (and even that we) obtain sustenance from the land. It is unconscionable that so many people do not know where their food comes from. Gardening and small-scale farming should be a mandatory part of every child's education.

Ultimately if we are to regain our health we need to relearn how to relate to nature. In the words of Joni Mitchell, we are "caught in the devil's bargain, and we've got to get ourselves back to the garden." This is no less true now than it was in 1969.

So, what we want to persist is nothing more than the soul of humanity—our *better* nature that relates, both cognitively and erotically via our animal senses, to the world. What can be done to ensure that it does, whatever else happens?

First we need to scale down. We will do that of course, one way or another, but it would be preferable to do it intentionally, nonviolently, and as soon as possible in an anticipatory response to what our best models are telling us. The impediment to this is our habitual dependency on others doing things *for* us, a perverse outcome of our system of 'education'. A healthier system would teach survival and self-reliance toward fulfillment of healthy needs. These would include the need for relationships, and the need for gratification that produces self-worth.

The emphasis on competition to the exclusion of cooperation has produced a perverse culture of celebrity worship that reinforces deference to fashion. In reality each of us can practically do anything, but each of us also has unique insights and creative talents that can be developed in a way that is fulfilling of both self and society. A healthy culture would take to heart the saying McLuhan attributed to the Balinese: "we have no art...we do everything as well as we can." Marx stated it as "from each according to his abilities to each according to his needs". These include the need for sustenance and love.

As we have argued, the Global Economy is based on highly questionable premises, a failure to think long-term (i.e., anticipate), and an unsustainable rate of energy flow. Our notion of wealth is misconceived, based on flawed materialist sense of value that does not account for the *actual* cost of living, or the inherent value of life itself. We cannot go back in time (nor would most of us want to), but we can 'go local'. It begins with acute awareness, which informs individual choices: in where we shop and what we buy for food, water, and shelter, and how much of a 'footprint' our lifestyle is making on the planet. Many of us are already consciously making such choices. But unfortunately, many more can't afford to because of 'market forces'.

But be wary of false advertising that promotes token gestures that make it easy to fool yourself into thinking that you are making things better, when in reality all you are doing is assuaging your own guilt as the system grinds on. Hypocrisy and *ex post facto* rationalization (i.e., self-deception) go hand in hand, and are a bane of our addiction to modern civilization.

Which brings us to politics and economics—two sides of the same coin. The system we have here and now in the United States (which is similarly modeled in much of the Western and increasingly Westernized world), in which the choice of politician is determined by economic and hence industrial power, is obviously not going to change itself, from the top-down. The only way that that people can change the system is through individual *resistance* and activism from the bottom up—a reality that participants of the Occupy Movement seem to have grasped. This will further stress the Global Economy, accelerating the collapse that we have argued is all but inevitable. But it will also build a foundation, hopefully while there is still time, for a new system to emerge from the ruins of the old. The objective here is to engender a metamorphosis that preserves the soul of humanity, and as healthy an environment as possible, preventing their demise in a maelstrom of war and famine.

What we need to work on then is building the foundation for a civilization that sustains, rather than degrades, the regenerative capacity of the biosphere. If we are going to create such a civilization we need to know what it is about life that affords regenerative capacity, and to assimilate that knowledge into cognitive frames that inform the collective unconscious. We (some of us at any rate) already know enough—in fact, it has been part of the Western scientific canon since Darwin, and before that it was encoded in 'folk wisdom' and the mythologies of many 'primitive' cultures. The basic idea, grasped by Darwin, is that adaptive resilience requires *copious undirected variation*. At the level of a species or population this means genetic diversity; that is why inbreeding is usually a fast track to extinction. At the level of ecosystems it means diversity of species and trophic 'flows'. In societies it means diversity of perspectives—'multiculturalism' (in a broad sense)—engendered by, and engendering, greater respect for and appreciation of the 'other'.

As discussed in Chapter 4, copious undirected variation in complex systems is a sign of immaturity, a relatively unconstrained stage of development. But to be effective at anything a system must develop a measure of maturity, manifesting as organized constraints that direct flow along specific paths. The question then is how much development is optimal? Here, as in most things, the Goldilocks principle applies. In life, health entails resilience, which is maximized by way of an optimal balance between undirected variation and developed constraint (Ulanowicz, 2009b). In a system, too much of either is unhealthy and ultimately fatal. Hence, to maximize resilience development must be regulated toward achieving the optimal balance.

Copious undirected variation, the raw material for evolutionary change, is what enables competition to work as a driver of improvement. But competition, widely touted as the primary source of all that is good, is in fact secondary (Ulanowicz, 2009a). From an ecological perspective it only works because cooperative mutualism among diverse components and networks of flows creates opportune niches that make it rewarding. A society (like ours) that values competition over cooperation loses touch with this reality, and thus tends to overdevelop a system of a few dominant flows, eroding the diversity that affords regenerative capacity, adaptive resilience, and the very niches that reward competition. A culture in which people are concerned with preserving the long-term regenerative capacity of civilization would not allow that to occur.

The question then becomes: how do we teach that? How do we teach the importance of making choices that maintain or increase ecological resilience, healthy balance, and long-term potential—choices that create options, rather than simply closing them off (Juarrero, 1999)? To answer this we might consider how many different ways the Modeling Relation can be used to obtain anticipatory knowledge. One way—the way of the physical sciences—uses mathematical formalism to make specific predictions about the behaviors of simple material systems. But as we have seen, this is an unrealistic approach to anticipating the behavior of complex systems, and hence of life itself. Complex systems are unpredictable in their specifics, and can't be controlled or adequately understood using formal models or algorithms. But that doesn't mean that we can't develop models that allow us to sustain regenerative capacity and adaptively anticipate the future. As animals we already do that to some extent unconsciously. What we need then is to bring that innate animal nature and ability back into public consciousness (Abram, 2010)—into art, science, and education—and augment it with better scientific models that recognize the generality, complexity, relationality, and subjectivity of life itself.

Our discussion of the logic of development in Chapter 4 is an attempt in that direction, building on the thinking of Charles Sanders Peirce, Robert Rosen, Stan Salthe, and Robert Ulanowicz, among others. The basic idea—that development is a teleological process of (self)- actualization via growth and self-organization, which when continued past a certain threshold of maturity causes senescence and thence the demise of the specific system or thing being developed—constitutes a model whose realization we are now witnessing at the level of global civilization. If that model were assimilated through teaching and learning and then it might afford better anticipatory responses—e.g., work toward conservation and establishment of regulations that prevent economic overdevelopment into senescence.

As discussed in Chapter 4, organisms that avoid developing into senescence are those that have the ability to both regulate their growth, and keep growing throughout their lifespan. For civilization to do that however (barring the Kurzweilian cyborganic scenario discussed above) would require that it revise its understanding of growth, in order to learn how to grow in ways that do not depend on extraction of non-renewable resources and environmental degradation. In organisms growth can be sustained only when it is coupled to recycling via deconstruction (self-digestion and cell death). Civilization may well be able to sustain some form of growth if it followed an analogous model, but it is difficult to imagine how that could be done morally. But perhaps 'growth' can be thought of in a more metaphorical sense and still be relevant to the problem. For example, the malaise experienced during aging is caused in no small part by cessation of psychological 'growth', giving rise to the aphorism 'you are only as old as you feel'.

So to create a civilization that realizes more adaptively resilient models, perhaps we need to embrace the idea that learning and education are a lifelong process. Moreover, science ought not to be pigeonholed as the occupation of specialists and 'experts'. At its best, it is an artful approach to life that embraces uncertainty and the creation of new models, one that happily discards models that are found not to fit with reality. At the same time, science should not be considered more important or valuable than other creative human endeavors, or even separate from them. So education should seek to break down (rather than reinforce) communication barriers that develop between different fields of inquiry. Ultimately, we need to recognize that *all* of our education and learning occurs by way of narrative—the telling and re-telling of stories, which become myths, which come to inform the collective unconscious and thus mold our behavior. Story-telling is central even to science, as every scientist who is honest with him or herself knows.

In sum, maintaining the regenerative capacity of civilization—its adaptive resilience and sustainability—requires new and better ways of teaching, learning, and relating to one another, ways that reground us in our right-brain mode of curious, vigilant attention that seeks to understand *the other*. This will not happen overnight, but it is beginning to happen now, particularly among young people. It appears to be part of the *modus operendi* of the Occupy Movement.

In the meantime it pays to also work, as much as possible (and as much as it goes against our better instincts), *within* the system to buy time. None of the choices that it offers are good—as we have noted, both 'sides' of the political divide are a part of, rather than a solution to, the problem. But as long as the system persists (and it will for a while, no matter what happens) it behooves us to try to work with it. All the choices may be bad, but some are better than others, and as they say, to choose not to choose is still a choice, with consequences. That is why, in what is passed off as 'democracy' in the United States, it is better to vote than not to vote—even if your choice amounts to is selecting a lesser of evils. 'Conservative' choices, motivated by a desire to restore traditional values and/or short-term quest for profit, will only make the transition more difficult to negotiate. 'Liberal' choices, such as they are, will not prevent the inevitable, but they might delay it a bit, and offer a somewhat better chance of alleviating the suffering through 'shared sacrifice'.

America received a hard lesson following the 2008 presidential election—a lesson that has by and large not yet been learned. The charismatic and intelligent Barack Obama was elected on a surging tidal wave of populist hope for change, fueled in no small part by disgust with the previous administration and the widespread feeling of powerlessness that its eight years engendered. Obama restored some feeling of 'power to the people'. But he also came into office at a bad time, when the economy was tanking and the nation was engaged in two wars that most people had little stake in and did not support. Human rights abuses had skyrocketed as a result of the terrifying "war on terror". Obama promised to fix those problems, and we believed him. But as they say, pie-crust promises are easily broken.

So many became disillusioned with the president, as the groundswell of public support that put him in office dissolved into righteous indignation among many. He wooed us in his campaign, but after getting into office his policies turned out not to be all that much different than those of his predecessor. So we the people who put him where he is felt used.

But it is important to view things in context and ask: how much is the man, and how much is the system? The context that allowed Roosevelt to do what he did in the 1930s no longer exists—the system has developed way past that. And even Roosevelt disillusioned many of his supporters. The fact of the matter is that no single human being, not even the president of the United States, is going to be able to change the system—the military-industrial-congressional complex that Eisenhower saw all too clearly and warned us about. Nevertheless, the president is extremely powerful, and that power can be used for great good or great ill. So we need to be very careful in making whatever electoral choices we are presented with, no matter how *bad* those choices are. It is good and extremely important to speak out against the current administration, but when it comes time to vote we need to carefully consider the consequences of not voting or voting for a nonviable candidate. This is a hard pill to swallow, made harder by the fact that the powers-that-be know it and count on it.

So working within the system (simply voting) is nowhere near enough. Every day that the Global Economy continues consuming resources creates significant losses both of and for future life. In addition to changing our selves we need to speak out, take to the streets, protest—exactly what Occupy Wall Street is now doing. This is the only way to counter the illusion that what is depicted on television, in the news, and in political speeches is (the only possible) reality.

At some point people need to realize that the opportunities for standard of living enjoyed by the middle class of a few decades ago are a thing of the past, and that chances of anyone realizing the 'American Dream' of accumulating large amounts of material 'wealth' are vanishingly small and far from equally distributed; that, in fact, we have developed to the point where a burgeoning middle class and accumulation of material wealth puts us on the fast track to ruin. The sooner people realize that the system is not working in their interests—including the interests of 'values voters'—the better.

In the meantime, the main reason to vote for liberal or 'progressive' politicians is that strong government regulations are needed to impede environmental degradation. If the right-wing corporatists have their way, regulations will be completely gutted in a last ditch effort to stimulate economic growth. This will only make things far worse than they already are, and consign our children and descendents to a hell on earth.

In conclusion, we are now confronted with a stark choice—basically, a decision of whether or not to walk a tight rope to safety—and not much time to make it. We can continue living in the past, working within an overdeveloped system to keep doing the same thing over and over and hoping for a different outcome—a course of action often touted as the 'definition of insanity'. Or we can work diligently to create a *new* system that does not degrade humanity, and build lifeboats for the harrowing transition that is nigh upon us.

In short, we can continue on in denial, and end up like the Titanic. Or, we can work in a concerted way at improvising a solution. Maybe we are dreamers in holding out hope that this is possible. But then so to were Jim Lovell, John Swigert, and Fred Haise, in holding out hope of safe return to earth. As their story shows, such dreams sometimes do come true. It starts with acute awareness—with the recognition that "Houston, we have a problem."

# REFERENCES

Abram, D. (2010). *Becoming Animal: An Earthly Cosmology*, ISBN 0375421718.

Agin, D. (2009). *More than Genes: What Science Can Tell Us about Toxic Chemicals, Development, and the Risk to Our Children*, ISBN 0195381505.

Agin, D. (2006). *Junk Science: An Overdue Indictment of Government, Industry, and Faith Groups that Twist Science for their Own Gain*, ISBN 0312352417.

Anderson, P. W. (1972). "More is different: broken symmetry and the hierarchical structure of science", *Science*, ISSN 0036-8075, 177: 393-396.

Arnold, D. A. (2002). *The Great Bust Ahead: The Greatest Depression in American and UK History is Just Several Short Years Away*, ISBN 159196153X.

Arthur, W. B. (1989). "Competing technologies, increasing returns, and lock-in by historical events", *The Economic Journal*, ISSN 0013-0133, 99: 116-131, http://www.jstor.org/stable/2234208.

Arthur, W. B. (2009). *The Nature of Technology: What it is and How it Evolves*, ISBN 1416544054.

Bejan, A. and Lorente, S. (2010). "The constructal law of design and evolution in nature", *Philosophical Transactions of the Royal Society B: Biological Sciences*, ISSN 1471-2970, 365: 1335-1347.

Bergmann, A. and Steller, H. (2010). "Apoptosis, stem cells, and tissue regeneration", *Science Signaling*, ISSN 1937-9145, 3: re8

Berry, W. (2000). *Life is a Miracle: an Essay against Modern Superstition*, ISBN 1582431418.

Berry, W. (2003). *Citizenship Papers*, ISBN 1593760000.

Buss, L.W. (1988). *The Evolution of Individuality*, ISBN 0691084696.

Campbell, J. with Moyers, B. (1991). *The Power of Myth*, ISBN 0385418868.

Carson, R. (1962). *Silent Spring*, ISBN 0618249060.

Cilliers, P (1998). *Complexity and Postmodernism: Understanding Complex Systems*, ISBN 0415152879.

Clarke, A. C. (1962). *Profiles of the Future: An Inquiry into the Limits of the Possible*, ISBN 0446321079.

Coffman, J. A. (2006). "Developmental ascendency: From bottom-up to top-down control", *Biological Theory*, ISSN 1555-5542, 1: 165-178, http://www.mitpressjournals.org/doi/pdf/10.1162/biot.2006.1.2.165.

Coffman, J. A. (2009). "On the subjective nature of reality, and its relationship to the objective reality of nature", *Philosophy Pathways*, ISSN 2043-0728, 144, http://www.philosophypathways.com/newsletter/issue144.html.

Coffman, J. A. (2011). "Information as a manifestation of development", *Information*, ISSN 2078-2489, 2(1): 102-116.

Coffman, J. A. and Mickulecky, D. C. (2012). "Of metaphors, metaphysics, and math: a mythology of mechanisms", *Philosophy Pathways*, ISSN 2043-0728, 170, http://www.philosophypathways.com/newsletter/issue170.html.

Crutzen, P. J., and Stoermer, E. F. (2000). "The 'Anthropocene'", *Global Change Newsletter*, ISSN 0284-5865, 41: 17-18.

Darwin, C. (1859). *On the Origin of Species by Means of Natural Selection, or The Preservation of Favoured Races in the Struggle for Life*, ISBN 0553214632.

Davidson, E. H. (2006). *The Regulatory Genome*, ISBN 0120885638.

Dawkins, R. (1976). *The Selfish Gene*. ISBN 0192860925 (1990).

Deacon, T. W. (2011). *Incomplete Nature: How Mind Emerged from Matter*, ISBN 0393049914.

Deloria, Vine Jr. (2003). *God Is Red*, ISBN 1555914985.

Diamond, J. (1987). "The worst mistake in the history of the human race", *Discover Magazine* ISSN 0274-7529, May 1987, pp. 64-66, published online May 1, 1999: http://discovermagazine.com/1987/may/02-the-worst-mistake-in-the-history-of-the-human-race/.

Diamond, J. (1997). *Guns, Germs, and Steel: The Fates of Human Societies*, ISBN 0393038912.

Diamond, J. (2004). *Collapse: How Societies Choose to Fail or Succeed*, ISBN 0670033375.

Durant, J. M., Hjermann, D. O., Ottersen, G., and Stenseth, N. C. (2007). "Climate and the match or mismatch between predator requirements and resource availability", *Climate Research*, ISSN 1616-1572, 33: 271-283.

Eagleton, T. (2011). *Why Marx Was Right*, ISBN 0300169434.

Ebert, T. A. (2008). "Longevity and lack of senescence in the red sea urchin *Strongylocentrotus franciscanus*", *Experimental Gerontology*, ISSN 0531-5565, 43: 734-738.

Eco, U. (2000). *Kant and the Platypus: Essays on Language and Cognition*, ISBN 009927695X.

Elsasser, W. M. (1972). "A model of biological indeterminacy", *Journal of Theoretical Biology*, ISSN 0022-5193, 36(3): 627-33.

Fernandez, E. (2008). "Biosemiotics and self-reference from Peirce to Rosen", Eighth Annual International Gatherings in Biosemiotics, University of the Aegean, Syros, Greece, June 23-28, http://www.lindahall.org/services/reference/papers/fernandez/PRfinal.pdf.

Finch, C. E. (1990). *Longevity, Senescence, and the Genome*, ISBN 0226248895.

Finch, C.E. (2009). "Update on slow aging and negligible senescence: A mini-review", *Gerontology*, ISSN 0304-324X, 55: 307-13.

Gould, S. J. (1989). *Wonderful Life: The Burgess Shale and the Nature of History*, ISBN 0393027058.

Gould, S. J. (2002). *The Structure of Evolutionary Theory*, ISBN 0674006135.

Hales, N. K. (1999). *How We Became Posthuman: Virtual Bodies in Cybernetics, Literature, and Informatics*, ISBN 0226321460.

Haken, H., Karlqvist, A, and Sveden, U. (1993). *The Machine as Metaphor and Tool*, ISBN 3540558160.

Harris, S, (2010). *The Moral Landscape: How Science Can Determine Human Values*, ISBN 1439171211.

Hedges, C. (2009). *Empire of Illusion, the End of Literacy and the Triumph of Spectacle*, ISBN 1568584377.

Heinberg, R. (2004). *Powerdown: Options and Actions for a Post-Carbon World*, ISBN 0865715106.

Kercel, S. W. (2007). "Entailment of ambiguity", *Chemistry and Biodiversity*, ISSN 1612-1880, 4(10): 2369-2385.

Juarrero, A. (1993). "From modern roods to postmodern rhizomes", *Diogenes*, ISSN 0392-1921, 41: 27-43.

Juarrero, A. (1999). *Dynamics in Action: Intentional Behavior as a Complex System*, ISBN 0262600471 (2002).

Jaynes, J. (1976). *The Origin of Consciousness in the Breakdown of the Bicameral Mind*, ISBN 0395207290.

Kelly, K. (2010). *What Technology Wants*, ISBN 0670022152.

Lakoff, G. (2008). *The Political Mind: Why You Can't Understand 21$^{st}$ Century American Politics with an 18$^{th}$ Century Brain*, ISBN 0670019275.

Lakoff, G. and Johnson, B. B. (2003). *Metaphors We Live By*, ISBN 0226468011.

Lakoff, G. and Johnson, B. B. (1999). *Philosophy in the Flesh: The Embodied Mind and its Challenge to Western Thought*, ISBN 0465056741.

Lakoff, G. and Nunez, R. E. (2000). *Where Mathematics Comes From: How the Embodied Mind Brings Mathematics into Being*, ISBN 0465037704.

Laland, K. N., Odling-Smee, F. J. and Feldman, M. W. (1999). "Evolutionary consequences of niche construction and their implications for ecology", *Proceedings of the National Academy of Sciences USA*, ISSN 0027-8424, 96 (18): 10242-10247.

Levin, J. (2006). *A Madman Dreams of Turing Machines*, ISBN 1400032407 (2007).

Levitis, D. A. (2011). "Before senescence: the evolutionary demography of ontogenesis", *Proceedings of the Royal Society B: Biological Sciences*, ISSN 1471-2954, 278(1707): 801-809.

Longo, G. and Montévil M. (2012). "The inert vs. the living state of matter: extended criticality, time geometry, anti-entropy - an overview", *Frontiers in Physiology*, ISSN 1664-042X, 3: 39.

Louie, A. H. (2007). "A Rosen etymology", *Chemistry and Biodiversity*, ISSN 1612-1880, 4(10): 2296-2314.

Louie, A. H. (2009). *More than Life Itself: A Synthetic Continuation in Relational Biology*, ISBN 3868380442.

Magdoff, F. and Foster, J. B. (2011). *What Every Environmentalist Needs to Know about Capitalism: A Citizen's guide to Capitalism and the Environment*, ISBN 1583672419.

Mailer, N . (1963). "The Big Bite", *Esquire,* April and May, pp. 159, 178-79, reprinted in *The Presidential Papers*, ISBN 0425018768.

Mailer, N. (1970). *Of a Fire on the Moon*, ISBN 0316544116.

Mann, M. E. (2012). *The Hockey Stick and the Climate Wars: Dispatches from the Front Lines*, ISBN 023115254X.

Margulis L. (1981). *Symbiosis in Cell Evolution*, ISBN 0716712563.

Mathews, F. (2003). *For Love of Matter: A Contemporary Panpsychism*, ISBN 0791458083.

McCarthy, C. (2006). *The Road*, ISBN 0307265439.

McGilchrist, I. (2009). *The Master and his Emissary: The Divided Brain and the Making of the Western World*, ISBN 0300168926 (2010).

McLuhan, M. (1962). *The Gutenberg Galaxy: The Making of Typographic Man*, ISBN 0802060412.

Meikle, J. L. (1997). "Material doubts: the consequences of plastic", *Environmental History*, ISSN 1930-8892, 2(3): 278-30.

Meyer, K. (2012). "Another remembered present", *Science*, ISSN 1095-9203, 335: 415-416.

Mikulecky, D. C. (1993). *Applications of Network Thermodynamics to Problems in Biomedical Engineering*, ISBN 0814754902.

Mikulecky, D. C. (2000). "Robert Rosen: the well-posed question and its answer – why are organisms different from machines?", *Systems Research and Behavioral Science*, ISSN 1099-1743, 17(5): 419-432.

Mikulecky, D. C. (2007a). "Complexity science as an aspect of the complexity of science", In C. Gershenson, D. Aerts, & B. Edmonds (eds.), *Worldviews Science and Us: Philosophy and Complexity*, ISBN 9812705481, pp. 30-52.

Mikulecky, D. C. (2007b). "Causality and complexity: the myth of objectivity in science", *Chemistry and Biodiversity*, ISSN 1612-1880, 4(10): 2480-2490.

Mikulecky, D. C. (2010). "A New approach to the Theory of Management: manage the real complex system, not its model", In S. E. Wallis (ed.), *Cybernetics and Systems Theory in Management: Tools, Views, and Advancements*, ISBN 161520668X.

Mikulecky, D. C. (2011). "Even more than Life Itself: beyond complexity", Axiomathes, ISSN 1572-8390, 21(3): 455-471.

Monod, J. (1972). *Chance and Necessity: An Essay on the Natural Philosophy of Modern Biology*, ISBN 0394718259.

Pattee, H. (2001). "The physics of symbols: bridging the epistemic cut", *Biosystems*, ISSN 0303-2647, 60: 5-21.

Powell, S. G. (2012). *Darwin's Unfinished Business: the Self-Organizing Intelligence of Nature*, ISBN 1594774404.

Reich, R. B. (2007). *Supercapitalism: The Transformation of Business, Democracy, and Everyday life*, ISBN 0307265617.

Rosen, R. (1958). "The representation of biological systems from the standpoint of the theory of categories", *Bulletin of Mathematical Biophysics*, ISSN 0007-4985, 20: 317-341.

Rosen, R (1972). "Some relational cell models: the metabolism – repair system", in R. Rosen (ed.), *Foundations of Mathematical Biology. Vol 2: Cellular Systems*, Academic Press, NY, pp. 217-253.

Rosen, R. (1973). "On the relation between structural and functional descriptions of biological systems", in M. Conrad and M.E. Magar (eds.) *The Physical Principles of Neuronal and Organismic Behavior: Proceedings*, ISBN 067712290X.

Rosen, R. (1975). "Biological Systems as paradigms for Adaptation", in R. H. Day and T. Groves, (eds.) *Adaptive Economic Models*. New York, NY: Academic Press.

Rosen, R. (1985). *Anticipatory Systems: Philosophical, Mathematical & Methodological Foundations*. ISBN 008031158X.

Rosen, R. (1986a). "Causal structures in brains and machines", *International Journal of General Systems*, ISSN 1563-5104, 12: 107-126.

Rosen, R. (1986b). "Some Comments on Systems and System Theory", *International Journal of General Systems*, ISSN 1563-5104, 13: 1-3.

Rosen, R. (1991). *Life Itself: A Comprehensive Inquiry into the Nature, Origin, and Fabrication of Life*, ISBN 0231075650 (2005).

Rosen, R. (2000). *Essays on Life Itself*, ISBN 0231105118.

Roszak, T. (1972). *Where the Wasteland Ends: Politics and Transcendence in Postindustrial Society*, ISBN 0385027389 (1973).

Sagan, D. (2007). *Notes from the Holocene: A Brief History of the Future*, ISBN 1933392320.

Salthe, S. N. (1993). *Development and Evolution: Complexity and Change in Biology*, ISBN 0262193353.

Salthe, S. N. (2003). "Infodynamics, a developmental framework for ecology/economics", *Conservation Ecology*, ISSN 1195-5449, 7(3): 3, http://www.consecol.org/vol7/iss3/art3/.

Salthe, S. N. (2009). "The system of interpretance, naturalizing meaning as finality", *Biosemiotics*, ISSN 1875-1350, 1: 285-294, http://dx.doi.org/10.1007/s12304-008-9023-3.

Salthe, S. N. (2010). "Development (and evolution) of the universe", *Foundations of Science*, ISSN 1572-8471, 15: 357-367.

Salthe, S. N. (2012). "Hierarchical structures", *Axiomathes*, ISSN 1572-8390, 22(3): 355-383.

Schneider, E. D. and Kay, J. J. (1994). "Life as a manifestation of the second law of thermodynamics", *Mathematical and Computer Modeling*, ISSN 0895-7177, 19: 25-48.

Schneider, E. D. and Sagan, D. (2005). *Into the Cool: Energy Flow, Thermodynamics, and Life*, ISBN 0226739368.

Schrödinger, E. (1944). *What is Life?*, ISBN 0521427088 (1992).

Shagrir, O. (2006). "Gödel on Turing on Computability", in A. Olszewski, J. Wolenski and R. Janusz (eds.), *Church's Thesis after 70 years*, ISBN 3938793090, pp. 393-419.

Singer, P. W. (2009). *Wired for War: The Robotics Revolution and Conflict in the 21$^{st}$ Century*, ISBN 1594201986.

Smith, E., and Morowitz, H. J. (2004). "Universality in intermediary metabolism", *Proceedings of the National Academy of Sciences USA*, ISSN 0027-8424, 101: 13168-13173.

Snow, C. P. (1998). *The Two Cultures*, ISBN 1107606144 (2012).

Soros, G. (2006). *The Age of Fallibility: Consequences of the War on Terror*, ISBN 1586483595.

Soros, G. (2008). "The Crisis & What to do about it", *The New York Review of Books*, ISSN 0028-7504, LV(19): 63-65.

Stager, C. (2011). *Deep Future: The Next 100,000 Years of Life on Earth*, ISBN 0312614624.

Thompson, E. (2007). *Mind in Life: Biology, Phenomenology, and the Sciences of Mind*. ISBN 0674057511 (2010).

Turing, A.M. (1936). "On computable numbers, with an application to the Entscheidungsproblem", *Proceedings of the London Mathematical Society*, ISSN 0024- 6093, 42: 230-265.

Ulanowicz, R. E. (1997). *Ecology: The Ascendent Perspective*, ISBN 023110829X.

Ulanowicz, R. E. (2009a). *A Third Window: Natural Life beyond Newton and Darwin*, ISBN 159947154X.

Ulanowicz, R. E. (2009b). "Increasing entropy: heat death or perpetual harmonies?", *International Journal of Design & Nature and Ecodynamics*, ISSN 1755-7445, 4(2): 83-96.

Volk, T. and Sagan, D. (2009). *Death and Sex*, ISBN 160358143X.

Wang, X. and Sommer, R. J. (2011). "Antagonism of LIN-17/Frizzled and LIN-18/Ryk in nematode vulva induction reveals evolutionary alterations in core developmental pathways", *PLoS Biology*, ISSN 1544-9173, 9(7): e1001110.

Wigner, E. (1960). "The unreasonable effectiveness of mathematics in the natural sciences", *Communications on Pure and Applied Mathematics*, ISSN 1097-0312, 1(1): 1-14.

Wolin, S. S. (2008). *Democracy, Incorporated: Managed Democracy and the Specter of Inverted Totalitarianism*, ISBN 0691135665.

Zimmerman, M. E. (2005). "Integral ecology: a perspectival, developmental, and coordinating approach to environmental problems." *World Futures: The Journal of General Evolution*, ISSN 0260-4027, 61: 50-62.

CPSIA information can be obtained at www.ICGtesting.com
Printed in the USA
BVOW062117241212

309060BV00003B/14/P